Novum Organum II

Novum Organum II

Going beyond the Scientific Research Model

Chris Edwards

ROWMAN & LITTLEFIELD
Lanham • Boulder • New York • Toronto • Plymouth, UK

Published by Rowman & Littlefield
4501 Forbes Boulevard, Suite 200, Lanham, Maryland 20706
www.rowman.com

10 Thornbury Road, Plymouth PL6 7PP, United Kingdom

British Library Cataloguing in Publication Information Available

Library of Congress Cataloging-in-Publication Data

Edwards, Chris, 1977– author.
Novum organum II : going beyond the scientific research model / Chris Edwards.
pages cm
Includes bibliographical references.
ISBN 978-1-4758-0999-2 (cloth : alk. paper)—ISBN 978-1-4758-1000-4 (pbk. : alk. paper)—ISBN 978-1-4758-1001-1 (electronic) 1. Teaching—Philosophy. 2. Bacon, Francis, 1561–1626. Novum organum. I. Title.
LB1025.3.E344 2014
371.102—dc23
2014003648

∞™ The paper used in this publication meets the minimum requirements of American National Standard for Information Sciences Permanence of Paper for Printed Library Materials, ANSI/NISO Z39.48-1992.

Printed in the United States of America

For Ben, who survived, and for Blake, who helped him

Contents

Acknowledgments

Thanks to Tom Koerner for his willingness to take on this project and for his help with the Latin terminology. Thanks to Michael Shermer of *Skeptic* magazine, who edited and published many of the papers that underpinned the philosophy here. Special thanks, also, to the missus for being a pretty good gal.

Introduction

Original work is inherently rebellious.

—Richard Rhodes, *The Making of the Atomic Bomb*

It is often possible to explain the ideas while leaving out most of the symbols.

—Ian Stewart, *Visions of Infinity: The Great Mathematical Problems*

I come to set forth the true way for the interpretation of nature.

—Francis Bacon, *Novum Organum*

In 1620, the Englishman Francis Bacon (1561–1626) published his masterwork *Novum Organum,* which roughly translates from the Latin to "*New Method.*" Bacon sought nothing less than a total overhaul of the way in which natural philosophers (as early scientists were known) viewed information and education. Bacon seemed appalled at the paucity of knowledge that human beings possessed about the natural world and blamed this mass ignorance on a stagnant educational philosophy that put an emphasis on the study of old knowledge rather than on the creation of new knowledge. In *Novum Organum,* Bacon proposed a grand epistemological schema whereby educated persons would be those trained in the means of knowledge production rather than in the study of the classics. Bacon encouraged practitioners of his new method to be like bees—to go out into the world in search of new things and then make useful things from the discoveries. In the modern era, researchers, the university system, and medicine have embraced Bacon's concept, although almost surely unknowingly, and his positions have served humanity well. Yet, new developments in theoretical physics and philosophy lead to the conclusion that *Novum Organum* might need, not revision, but something that would leave the original intact while adding something to it: a sequel.

Before beginning, a few words about Bacon himself and then a few chapters about his philosophy. Having been born into a well-to-do family in 1561, his early career as a lawyer and politician played out under the reign of Queen Elizabeth I. Elizabeth, known by her, perhaps rather optimistic, nickname of the "virgin queen" for her refusal to take a husband, never favored Bacon with any important positions, and he stayed in the House of Commons.

When the childless queen died, James I (whose name would become immortalized in the King James Bible and in the settlement of Jamestown in the same way that Elizabeth's nickname remains with us in the form of Virginia) took the throne and favored Bacon with a series of high appointments and titles, eventually making him a baron.

Two avenues were open for advancement in the court of King James. One could either (a) be a handsome young man or (b) handle administrative work so that James could spend more time with handsome young men. Bacon fell into the latter category and at one point more or less governed England as a regent while the king was out of country.

Behold a man torn in two by ambition. Bacon wanted nothing less than to rise in political influence while at the same time creating a philosophical revolution without precedent in the history of ideas. He did both, and in the month of October in 1620 presented his monarch with *Novum Organum*. James wrote a kind letter of acceptance. Someone, however, later recorded a comment by the sovereign that indicates that Bacon's philosophy flew right over the red head of James.

Within a year of this monumental publication, Bacon found himself put on trial by Parliament. He'd taken his king's side over the issue of establishing monopolies and faced charges, potentially trumped up, of taking bribes from what might today be termed lobbyists. Found guilty, Bacon gave up his position and took on a life of forced retirement. Now in his sixties, he never seemed to have lost the energy of his youth. Like Machiavelli, who also philosophized after being exiled from politics, Bacon found plenty of philosophical illumination in his twilight years and continued to write.

Bacon's death sits in scientific history's cabinet of curiosities, along with Tycho Brahe's prosthetic silver nose, the streaking incident of Archimedes, and Isaac Newton's penchant for playing with (and perhaps drinking) mercury. In fact, a lesser writer might write that Bacon died with egg on his face.

In 1626, while conducting an experiment on the effects of cold as a food preservative, Bacon stuffed a chicken with snow and as a result caught pneumonia and died. At least this is the story. Cold does not cause pneumonia, of course, and the line of causation drawn between the frigid chicken and Bacon's demise seems a bit shaky upon deep inspection, but it makes for a good enough legend.

Bacon's genius was to be the first to recognize and understand what the new method of experimentation actually meant. Bacon saw, before anyone else, that the scattershot methods of experimentation represented a potential intellectual revolution if only those methods could be incorporated into a larger epistemological framework and put to use as the "new method" that all thinkers employed.

In other words, he saw that an experiment was, in fact, a way to produce new knowledge and that new knowledge improved our ability to understand and manipulate our environment. Each individual experiment represented a smaller manifestation of a larger shift in thinking, but no one before Bacon had yet articulated that larger shift.

Bacon did not disparage previous knowledge or thinkers, but simply didn't think that humanity should have looked at the collection of Greek, Roman, Christian, and Renaissance classics and declared, "Well, this is good enough." It was time to forcefully add to the database of knowledge in the world, and the scientific method provided the means to do so.

Likewise, this book seeks not to take away from Bacon's method. Indeed, readers can probably deduce my deep affection for the man and his ideas from this introduction. In fact, if asked, I would probably rank Bacon's book as the most important work in the history not only of science but of ideas. The entire modern scientific enterprise, including medicine, grows from Bacon's book, as does the modern concept of the university.

Yet it's time for something new again.

This book is based upon a series of papers that I wrote for *Skeptic Magazine* and some well-respected educational journals. I have incorporated many of these ideas in three books of philosophy and education. In my last book, *Teaching Genius: Redefining Education with Lessons from Science and Philosophy*, I attempted to connect the *new* new method of thought with an educational philosophy. This was, perhaps, a bit hasty of me. The *new* new method has been understood by only a few and accepted by fewer.

The attempt here is to compile many of the ideas and positions that come from my previous papers into a single, sequential, and logically ordered volume. I've already written about the implications of the *new* new method for education and will reiterate a few of those points here, but mostly I seek to explain the new methods for a general audience. It is my hope that, once this new method of thought reaches a larger audience, the logical sense inherent in my educational positions will be understood and more widely adopted.

Two previously published *Skeptic* articles are slightly modified and included in part II. My hope is to provide a larger context for those papers and to explain their further implications.

This is not a long book, but neither was the original *Novum Organum*. Bacon's clear prose, direct writing style, and clever use of analogies make him a continuing pleasure to read (as opposed to, say, *everybody else* who wrote philosophy at the time). Too many philosophers, then and now, write in figure eights and loop-de-loops. Like Bacon, I prefer to write in a straight line.

I

The *Old* New Method

Metacognition involves questioning the very parameters of the field upon which one is thinking. Thinking in such a way can be hard, since the boundaries of a field can be so ever-present as to be invisible. The study of history opens the mind to the concept of metacognition. Time and again, the student of history encounters previous thinkers banging their heads against walls they can't even see. We might chuckle at medieval scholastics, unable to question Aristotle, or at pre-Darwinian naturalists, unable to break away from the concept of a celestial Designer, but those chuckles can quickly become uncomfortable laughter when one realizes that modern thinkers operate under certain unquestioned assumptions too.

Look at the word "research." Search, and then, search again, it says. Every field involves the search for extra knowledge via experimentation. Experts in medicine are either expected to perform research via clinical trials or to remain expert in research findings. The social sciences are reliant upon experiments that provide new knowledge that helps to reveal the working of the human mind either in isolation or in group settings. Education, even, is reliant on the research model. Researchers try to make their samples as representative as possible of various groups so that practitioners can use the research information in analogous situations. We search and search and search and search and then re-search.

Francis Bacon created this scientific research model. It is important to state this so that we may remember that the scientific research model, like all works of scholarship or religion, is of human origin. To note the origins of something is to strip it of its mystery, and the stripping of mystery is the first step in critical analysis.

Bacon's name should be as well-known as that of any religious prophet and better known than that of many political leaders or social crusaders. Nobody ever wrote or said anything more influential than the words that Bacon put down in *Novum Organum*.

Take a moment to look at the books in the world history section of your local bookstore or library (before either, or both, disappears) and count the number of books with the dreadful subtitle of the "*blank* that changed the world." Many good books have been smeared by this least imaginative of addendums. After all, everyone and everything changes the world in some way, do they not?

Yet, Bacon really did change the world. I mean, he changed the *whole* world. He systematized the scientific method. In time, his new method would be adopted by every field in the West, and the rest of the world had to react. When people talk, and they still do, of China or the Middle East needing to westernize, what they really mean is that those societies need to break away from the rusty chains of tradition and adopt Bacon's new method of using scientifically based research to change society for the better.

But Bacon's new method has its limits. And before we can get beyond the scientific research model, we must demystify it. We have to see the boundaries of the model before we can see what lies beyond them.

ONE

The First *Novum Organum*

> Human knowledge, as we have it, is a mere medley and ill-digested mass, made up of much credulity and much accident, and also of the childish notions which we at first imbibed.
>
> —Francis Bacon, *Novum Organum*

Novum Organum was actually an addendum of a larger work titled the *Instauratio Magna*, or "*The Great Instauration.*" *Instauratio*, a Latin term, translates to "renewal" or "restoration." Interestingly, this restoration took place just slightly one hundred years after the Reformation, and the disrespect for authority that animated the latter likely affected the former. Bacon, in 1620, writes with occasional contempt for ancient authorities and, while he throws alms toward religious authorities, in fact proposes a rather radical break between religion and science.

Bacon can speak for himself. In the preface to *The Great Instauration* he writes:

> For let a man look carefully into all that variety of books with which the arts and sciences abound, he will find everywhere endless repetitions of the same thing, varying in the method of treatment, but not new in substance, insomuch that the whole stock, numerous as it appears at first view, proves on examination to be but scanty. And for its value and utility it must be plainly avowed that that wisdom which we have derived principally from the Greeks is but like the boyhood of knowledge, and has the characteristic property of boys: it can talk, but it cannot generate; for it is fruitful of controversies but barren of works. (p. 7)

Shortly thereafter, he writes:

> In the mechanical arts we do not find it so: they, on the contrary, as having in them some breath of life, are continually growing and becoming more perfect. As originally invented they are commonly rude,

> clumsy, and shapeless; afterwards they acquire new powers and more commodious arrangements and constructions; in so far that men shall sooner leave the study and pursuit of them and turn to something else, than they arrive at the ultimate perfection of which they are capable. Philosophy and the intellectual sciences, on the contrary, stand like statues, worshiped and celebrated, but not moved or advanced. (p. 8)

In the first of the preceding extracts, Bacon disparages not only a classical education, but also the entire realm of secondary scholarship based upon it. Please remember, as will be outlined here momentarily, that medieval Catholic scholars all but worshiped Aristotle and that Renaissance thinkers, especially Petrarch (who coined the term Renaissance), considered Greek philosophy, literature, and mathematics to be the basis for a "rebirth" in thought.

One modern scholar still credits the reemergence of Lucretius's classical-era secular poems and philosophies as being the "swerve" that sent the intellectual world hurtling toward a scientific worldview. Yet here was Bacon comparing all that the Greeks produced to the ramblings of prepubescent boys.

For those of us who still see value in pondering the Socratic dialogues, grimacing at the good parts of Herodotus, trying not to snooze through Thucydides, and wrestling with Aristotle's *Metaphysics*, or even chuckling through *Lysistrata*, such a disparagement of the classics seems rude at best, anti-intellectual at worst. But we live in a time where the degradation of a classical education is complete and where schools have pitched Socrates in favor of dry textbooks, or worse, interactive educational games.

Bacon lived in a world where intellectuals prized the Greeks overmuch and where educated peoples took more pride in their ability to read Greek or memorize lines of the classics than they did in their ability to produce useful ideas. Bacon put force in his punches because his opponent at that time stood upright, and if it seems a bit harsh today, that's because his former opponent now lays prostrate.

Specifically, Bacon seemed troubled by the attempt of philosophers, Greek, natural, or otherwise, to see some minor principle from a few examples and then try to extrapolate those principles into a general theory of the universe. He writes:

> There is also another class of philosophers, who having bestowed much diligent and careful labor on a few experiments, have thence made bold to educe and construct systems; wresting all other facts in a strange fashion to conformity therewith. (p. 46)

He especially went after Aristotle by stating:

> Men become attached to certain particular sciences and speculations, either because they fancy themselves the authors and inventors thereof, or because they have bestowed the greatest pains upon them and be-

> come most habituated to them. But men of this kind, if they betake themselves to philosophy and contemplations of a general character, distort and color them in obedience to their former fancies; a thing especially to be noticed in Aristotle, who made his natural philosophy a mere bondservant to his logic, thereby rendering it contentious and well-nigh useless. (pp. 41–42)

The direct criticism of the esteemed Aristotle itself deserves further commentary, and will shortly get it, but let's stick with Bacon's point for a moment. He's directly criticizing Aristotle for creating a logical system and then assuming without evidence that the phenomena will act in accordance with his systems of thought. Hypotheses are not supposed to bully the evidence, but instead should arise from the evidence itself.

Dogma comes from assuming evidence to fit prestructured philosophy, and science arises from using evidence to craft philosophy. In the same breath, Bacon criticizes William Gilbert, author of the classical work on magnetism called *De Magnete*; for he studied only the magnetized lodestone and then "proceeded at once to construct an entire system in accordance with his favorite subject" (p. 42).

Later, after abstracting the philosophies of the pre-Socratics—almost all of whom had a pet substance, be it water, air, fire, or indivisible particles called atoms, that they believed everything in the universe to be made of—Bacon again attacked Aristotle with this passage:

> In the physics of Aristotle you hear hardly anything but the words of logic; which in his metaphysics also, under a more imposing name, and more forsooth as a realist than a nominalist, he has handled over again. Nor let any weight be given to the fact that in his book on animals, and his Problems, and other of his treatises, there is frequent dealing with experiments. For he had come to his conclusion before: he did not consult experience, as he should have done, in order to the framing of his decision and axioms; but having first determined the question according to his will, he then resorts to experience, and bending her into conformity with his placets leads her about like a captive in a procession: so that even on this count he is more guilty than his modern followers, the schoolmen, who have abandoned experience altogether. (p. 47)

A paragraph later, Bacon cautions against studying a small sample and then jumping to universal truths. He specifically attacks Pythagoras as being a "striking example" of this school of thought. Indeed he was. Pythagoras, born about 570 BCE, believed the universe to be constructed of whole numbers. He reasoned to this conclusion by asking why certain combinations of music sounded melodic while other combinations, like what teenagers play in garages, sound so, well, unmelodic.

He concluded that ear-pleasing music resulted when notes were played in whole-number combinations. The triangle, with a structure that could be built from laying three rocks on top of four, and two on top of three, and one on top of two, also seemed to be built of whole numbers. From this, Pythagoras believed that he possessed enough information to assume the rest of the universe operated so neatly.

A highly entertaining PBS production (*The Story of 1*, featuring Terry Jones of *Monty Python* fame) about the history of the number one shows an actor playing Pythagoras losing his sanity as he realized that a ninety-degree triangle with two equal sides cannot be constructed out of units that are the same size; in other words, not everything could be explained through the "whole number" concept. (The video also drops the fun fact that Pythagoras abstained from beans, a peculiarity he also required of his followers, based on the belief that a fart represented the loss of a bit of one's soul.)

Now, these cited passages demonstrate the agility of Bacon's intellect. Not only did he chide Aristotle and the "schoolmen," also known as Scholastics, for their irresponsible use of deductive reasoning, but he also scolded Gilbert and Pythagoras for their misguided use of inductive reasoning.

Deductive reasoning uses general notions to describe a specific and takes the form of Aristotle's famous syllogisms. Everyone at some point has heard something to the effect of: "All politicians are corrupt. Dave is a politician; therefore, Dave is corrupt." We start with an "all" phrase, in this case about politicians, and then assume a particular instance, in this case "Dave," fits under the banner.

Inductive reasoning works in the opposite way. Someone studies a specific and then, based upon the evidence, moves to a general and broader idea. In his criticisms of induction, Bacon grasped the problem of sample size, which modern statisticians have since worked into a science.

In effect, Bacon criticized those who practiced induction for failing to have a large enough sample size and, therefore, falling into what we would today call a sampling error, which comes about as a result of not checking into a body (be it a pond or human population) widely enough to draw large-scale conclusions.

So important is a dose of humility to inductive reasoning that twentieth-century philosopher of science Karl Popper worked out the idea that no scientific hypothesis can absolutely predict the future. Therefore, a hypothesis can never be proven at all, only disproven.

You can see the damage inductive reasoning has wrought on dogma. It is remarkable, however, that it took over three hundred years between Bacon and Popper for the logic of inductive reasoning to reach its final conclusion. This will, however, not be the last time in this book that we encounter such a phenomenon.

Bacon engages his most severe language against Aristotle with this paragraph:

> For the philosophy of Aristotle, after having by hostile confutations destroyed all the rest (as the Ottomans serve their brothers), has laid down the law on all points: which done, he proceeds to himself to raise new questions of his own suggestion, and dispose of them likewise; so that nothing may remain that is not certain and decided,—a practice which holds and is in use among his successors. (p. 50)

The Ottomans, it should be noted, tended not to "serve their brothers" with any degree of kindness, being notable for their brutal expansionist policies. The Ottomans were founding members in the conquer-butcher-and-enslave club of late medieval empire builders. Bacon's analogy is that Aristotle's philosophical followers, the (in)famous Scholastics, served only to quash the free inquiry that animated the scientific "new method."

Bacon's acerbic tone toward Aristotle (in an era where the ancient Greeks' philosophy held sway over the university system in a similar way that the Koran dominates fundamentalist Islamic schools in the Middle East today) may not be the most remarkable aspect of Bacon's work. He proposed nothing less than the secularization of knowledge:

> Nothing is so mischievous as the apotheosis of error; and it is a very plague of the understanding for vanity to become the object of veneration. Yet in this vanity some of the moderns have with extreme levity indulged so far as to attempt to found a system of natural philosophy on the first chapters of Genesis, on the book of Job, and other parts of the sacred writings; seeking for the dead among the living: which also makes the inhibition and repression of it the more important, because from this unwholesome mixture of things human and divine there arises not only a fantastic philosophy but also an heretical religion. Very meet it is therefore that we be sober-minded, and give to faith that only which is faith's. (p. 48)

Give to faith that only which is faith's. The idea that the natural world should be studied through methods of science, and that faith had no place in this type of study, emanates from *Novum Organum*. For those paying attention, Bacon placed the philosophical mistakes of Aristotle and Pythagoras (exemplars of deduction and induction, respectively) in the same category as the errors made by those attempting to mix religion with natural explanations. How shocking is it, 260 years before Darwin, to read of a man critical of treating the book of Genesis as a book of natural science?

Having addressed the errors of philosophy, Bacon takes on the tone of a prophet when he writes, "I come to set forth the true way for the interpretation of nature" (p. 52).

There's arrogance, perhaps, in these words, but not hubris. Bacon largely accomplished his stated intent.

KEY POINTS

Francis Bacon criticized historical practitioners of both inductive and deductive reasoning for overstating their cases. His frustration with the seventeenth-century academic establishment stemmed from his belief that academics and philosophers should not be content to merely study the classics.

In *Novum Organum* Bacon singles out Aristotle for criticism. Bacon made a point to release the hounds of reason on Aristotle because of the worshipful way in which the "schoolmen" (Scholastics) of the early modern era treated Aristotle's ancient philosophy. Bacon believed this excess of respect prevented philosophers from creating new ideas.

Radically, for his time, Bacon argued for a break between scientific reasoning and religious faith, going so far as to argue that the book of Genesis should not be read as a literal truth. He believed that the attempt to reconcile faith and reason hurt both philosophy and religion.

TWO

The First Part of *Novum Organum* in Context

Since faith alone suffices for salvation, I have need of nothing, except that faith exercise the power and dominion of its own liberty.

—Martin Luther, *The Freedom of a Christian*

The discovery of evolution was held back by the dead hand of Plato.

—Richard Dawkins, *The Greatest Show on Earth: The Evidence for Evolution*

The quoted passages in the previous chapter, again, probably fail to shock the modern reader. Few people take offense over insults aimed at Aristotle, and the man himself remains thoroughly dead and, therefore, immune to hurt feelings. But to the late medieval world, to criticize Aristotle meant to criticize the entirety of a knowledge system that, errors and all, the intellectual world found comfort in. Even Aristotle must be put in his proper context, and to do that a brief foray into philosophical history is necessary.

Although the pre-Socratics said much of great interest and importance about the nature of the physical world, Aristotle and Bacon must be described in terms of their relationship to doubt. That relationship, after all, separates philosophy from theology and science from dogma. A brief philosophical history will help to illuminate the importance of Bacon's words. Western philosophy begins, in earnest, with Socrates.

So far as we know, Socrates wrote nothing during his philosophical tenure. Everything we know about the great philosopher comes from the works of two authors: Xenophon and Plato. A friend of Socrates, Aristophanes, referred to the philosopher in his comedic plays and even created one completely about Socrates called *The Clouds*, but these make no claim to historical accuracy. Plato's works on Socrates consist of several dialogues between Socrates and various people, sometimes students and sometimes philosophical foes and foils.

While it is unclear whether Plato's Socrates represents a real attempt to capture the historical man or whether Plato chooses to use Socrates as a fictional or semifictional character as a vessel for passing on Plato's ideas, the books themselves compose the cornerstone of a classical education.

In the books *Protagoras*, *Gorgias*, and *Meno*, Socrates expounds his philosophy through ruthless dialect, thus providing a stage for the Socratic method. The method involves responding to questions or assertions with more questions. At first this might seem almost juvenile, and can be if attempted by someone unskilled in philosophy, but when properly understood the methodology represents the pinnacle of the learning experience between student and teacher.

To begin, the teacher must know how to reason.

When he has reasoned to a conclusion he is sure of, he then merely invites his student to ask the same questions. Done properly, the student will follow each logical step until he ends up standing on the rhetorical hilltop, on the exact same conclusion as his teacher.

But the most famous and the most important of Plato's dialogues on Socrates involve his trial and execution by mandated suicide. These four works (the *Euthyphro*, *Apology*, *Crito*, and *Phaedo*) detail the best known drama in philosophical history.

By 399 BCE, Socrates had reached seventy. He was a veteran of the Peloponnesian War and lacked a proper occupation. The upstanding citizens of Athens apparently lost patience with this aged, notoriously ugly, and homeless atheist who reveled in asking questions designed to call into question the very existence of knowledge itself. They placed him on trial. Perhaps it was his charisma, or the fact that many youth show an inclination toward counterculture figures, but Socrates had no shortage of students.

In *Euthyphro*, Socrates has been indicted by a man named Meletus (the dialogue indicates that Socrates and his accuser had yet to meet) and is casually making his way to the judicial proceeding in Athens. The reader learns of this through Socrates's conversation with Euthyphro, a young lawyer whom the old philosopher encountered on his trip. Socrates reveals that Meletus "says he knows how our young men are corrupted and who corrupts them." What the corruption refers to has been well argued by classicists over the years.

The most common interpretation has it that Socrates, an atheist, had been questioning the existence of the gods. The scholar I. F. Stone (1989) counters this by noting that Socrates publicly criticized Athenian democracy while praising the Spartan system, and this at a time when the Athenians had just lost the brutal Peloponnesian War and endured the freedom-shattering rule of a Spartan-appointed oligarchy known as the Thirty.

Socrates's cause was not helped by the fact that the shenanigans of one his most famous pupils, Alcibiades, had helped the Spartans to win the war. Another, Critias, had tyrannical tendencies but died before the famous trial of his teacher.

Anybody with a mother who raised him right knows that two topics always stir up trouble and are therefore to be kept out of polite conversation: religion and politics. Socrates refused to stop talking about either, and his questions threatened to cut the already fragile social threads holding Athenian society together.

In *Euthyphro*, Socrates spends very little time discussing his own circumstances and instead noses his way into his young walking companion's business.

Euthyphro, it turns out, came across Socrates while on the way to his own legal proceeding. He will star as a prosecutor pursuing a murder case against his own father. This lawyer seemed to believe his actions to be virtuous, until Socrates requires that his companion define "virtue," at which point the young lawyer finds that the old man has him in an epistemological headlock.

Socrates verbally shakes the man with question after question until, finally, Euthyphro, no doubt speaking for thousands of new philosophy students over the years, states, "But Socrates, I have no way of telling you what I have in mind, for whatever proposition we put forward goes around and refuses to stay put where we establish it." After a long time, Euthyphro finally breaks free not with a superior logical argument but by crying the equivalent of "uncle" and saying, "I am in a hurry now, and it is time for me to go."

This dialogue contains all things Socratic: calmness in the face of persecution, an overwhelming interest in finding the truth, philosophical questioning/harassment, and finally a tacit admission that nothing can be truly known. Socrates loved to ask people to take the same logical steps that he did, only to lead them into a metaphysical wilderness known as a paradox. He may not have wanted everyone to live there with him, but he wanted to make sure they knew it existed.

Three men, Anytus, Lycon, and Meletus bring the charges against Socrates, and the old man goes before an Athenian jury of five hundred. *The Apology* features Socrates doing everything but apologizing. Instead, he bats Meletus around with rhetorical ease and in the process showcases the central tenets of philosophical inquiry, which can be defined as the process of taking the argument used in one case to see if it applies to other similar cases.

Meletus, mostly, accuses Socrates of atheism, of disbelieving in the gods. Furthermore, Socrates faces the charge of teaching his methods to others and making the weaker argument seem stronger! Socrates, typically, replies that he was no teacher and merely spends his time expounding on topics. If people within earshot hear him and ask questions or make comments, then how can he be to blame?

The whole *Apology* features as a rhetorical masterpiece, but the jury finds Socrates guilty anyway, by a vote of 220 to 280. The only show of emotion that the philosopher allows himself is to state that the verdict only came from the high esteem in which the jurors held Anytus and Lycon, and not because of the wretched rhetoric of Meletus. Socrates then famously states that "the unexamined life is not worth living for a man."

When asked what his punishment should be Socrates suggests a fine. The jury sentences him to death, which in Athenian society occurred by the drinking of a poison called hemlock. Unconcerned, Socrates merely notes that age would do the job just as well if the jurors can wait. Death brought either eternal sleep, in which case he would rest, or it brought an eternal change; neither prospect was to be feared.

I. F. Stone (1989), in particular, has written that the whole scene makes little sense in the context of typical Greek life. Playwrights had been joking about Socrates and the gods for years, and the polytheistic Athenians hardly appear, from the historical record, to have been religious fanatics. Maybe Socrates had not changed, but Athenian society, exhausted by the Peloponnesian War and the tyranny of the Thirty had lost its sense of humor. Socrates, the opponent of everything, no longer seemed so harmless.

In *Phaedo*, Socrates meets his death. This dialogue showcases the calm approach of the ethical man toward dying. In the *Apology*, Socrates states that nothing can harm a good man, and his conduct at the end of his life seems to uphold his philosophy. His students try to talk their beloved teacher into running away, as he is not seriously guarded. Socrates responds by spinning their arguments around, probing as to whether such an action could be ethically justified. One must comply with the laws, Socrates asserts, and when the time comes he takes the cup of hemlock and drinks it down while the surrounding students grieve.

And yet he lives on.

Not present at the death of Socrates was one his finest students, Aristocles (428–347 BCE). Because he was a former wrestler, everyone referred to him by his nickname of "broad" or "broad shouldered," which in Greek translated to "Plato." With his teacher gone, Plato abandoned Athens and then, after travelling to no one knows where, returned to establish the Western world's first university, called the Academy.

The ideas displayed in Plato's voluminous writings cannot be given justice in summary, but he is best known for the Socratic dialogues described above. His work on governmental theory, divided into ten books, is called *The Republic*, and through these works Plato argued that society should be governed by "philosopher kings" trained specifically in the art of reason. Plato distrusted the passion-enthused masses and, therefore, democracy. Reason should be the major driving force in decision-making, and, therefore, those who had mastered the art of reason should be in political control.

The single idea most attributed to Plato involves the analogy of the cave. In the seventh volume of *The Republic*, Plato compares those unfamiliar with philosophy to people chained to the floor in a cave. Imagine a fire burning behind them so that shadows form on the wall. If these prisoners never saw anything but the shadows, they would mistake these shades of reality for what is actually real.

This sets the mind up for the understanding of Plato's complex philosophy, which purports to deal with unseen realms that are more real than the everyday reality experienced by people. Think of a triangle: we all learn that triangles can be dissected and understood by certain geometric theorems. The same is true of circles or any other shape.

We mathematically work with perfect shapes. But then, try to actually draw a perfect right-angle triangle or mathematically precise circle, and you will see that, while you have some concept of what a precise geometric shape would look like, actually making such a shape appear in the tangible world is impossible.

So what is more real, the image of the shape and the mathematical concepts derived from it or the actual shape itself drawn before you? To Plato, the perfect circle or triangle must exist in another realm, and all of the triangles and circles that appear in this world are but imperfect shadows of the perfection.

Things are judged to be beautiful, or just, or ethical based upon how closely they match the real object in the other realm. Things are judged to be ugly, unjust, or evil depending upon how far away from the perfect object they are. (Single gentlemen might try this Platonic pick-up line: "Miss, you are as close to the Platonic ideal of the female as any woman can be.")

We can see now that Plato's philosophy amounts to a linguistic game. We have definitions, and real objects sometimes meet those definitions and sometimes do not. Plato had no understanding of how things might evolve. One can see why Plato's philosophy will be so endearing to Christians later.

The concept of this world being an imperfect shade of a perfect world and of people in this world being imperfect copies made in the image of a perfect creator fit nicely with embryonic Christian theology. So prevalent was the Platonic epistemological schema that Richard Dawkins, among others, believes that Plato's philosophy effectively blinded pre-Darwinian thinkers to evolution's bottom-up approach to describing biological objects.

Plato's student Aristotle disagreed, and, in some ways, all philosophical and theological arguments to be developed in the future have their roots in this argument between the philosophical titans of Greece.

Aristotle entered Plato's Academy at the age of seventeen. Coming from a wealthy family, Aristotle did not have to worry about making a living, and so he spent the next twenty years learning and teaching. When Plato died in 347 BCE, Aristotle moved on.

Acting as perhaps the first polymath, Aristotle pursued learning in all the major philosophical regions, delving into literary theory, science, rhetoric, physics, and formal logic. He invented deductive logic, where one creates theories based upon the evidence and also the syllogism. The syllogism, a formal logical construct often shown in the form of if A = B and B = C, then A = C, gave codified rules to thought.

The physics of Aristotle invoked an original "unmoved mover" or "prime mover," something that pushed the universe into being. Later, Muslims and Christians, enamored of Aristotle to begin with, would call this mover "Allah" or "God." The reasoning was simple: there had to be some first cause that pushed movement and the universe into being. (Aristotle's reasoning is not very convincing, as it does not answer the question of what the first cause was but rather introduces another question, such as, "From where did the Prime Mover come from?")

Aristotle disagreed with Plato as to the realm of forms and was in fact a materialist. This meant that he believed everything in the universe had to be made of some material. Tell Aristotle that you saw a ghost, and he likely would have replied that, if you saw the ghost, it must be made of something that light can bounce off of, or else it is self-luminescent. Either way, the ghost must be made out of some kind of tangible material.

He got a few things wrong, writing that the objects in the solar system revolved around the earth and creating a logical construct about motion that failed, eventually, to hold up to experimental scrutiny. He also tutored Alexander the Great, but seemed to have little impact. Aristotle was a great thinker, while Alexander had other interests, such as the conquest of the known world, in mind.

Aristotle's philosophy continued to be taught until the sixth-century Roman emperor (residing in the eastern half of Rome known as Byzantium) Justinian exiled the teachers of Aristotle's philosophy. It's not clear why, except that his contemporaries seemed to think Justinian to be a bit feeble-minded and a Christian in addition, and neither group tended to think much of philosophers.

Many of the philosophers fled to Syria, where they just happened to be when the Arab/Muslims conquered the territory, along with just about everywhere else, in the early seventh century. In the eighth century, with the rise of the second great Islamic dynasty, that of the Abbasid, a series of philosophically minded rulers facilitated a type of Islam that treated freethinking as a version of worship.

The caliph Al-Mamun took the position, known as the *Mu'tazili* thesis, that the Koran represented a human-made creation subject to the interpretation of the caliphate. This governmental position created an intellectual environment that facilitated not only the building of a great library but also the development of a philosophical approach to Islam.

The Abbasids moved the capital of the caliphate from Damascus, in Syria, to Baghdad, in Iraq, and here, at the newly built library known as the House of Wisdom, Jewish, Christian, and later Islamic scholars began a process of translating the world's knowledge into Arabic. The scholars probably put no more emphasis on their religious distinctions than doctors in a modern hospital or scholars in a modern university would.

Here at the House of Wisdom, Aristotle found new life. A few of the translators discovered his work and began to try to integrate it with the faith-based concepts of Islam. This seemed not to work, as Aristotle's forms of reasoning caused friction with faith. One Islamic scholar known in the West as Avicenna (980–1037) worked Aristotle's philosophy into his own work. In her book *Doubt: A History*, Jennifer Michael Hecht (2003) summarizes Avicenna's positions:

> As far as [Avicenna] was concerned, anyone who was capable of philosophy was called to do it, and would want to—it was a moral responsibility to seek truth, but it was also the best game in town. God was understandable intellectually, and that was the sweetest and highest way. Intellectuals thus had access to some of the joys of revealed religion without entirely contradicting their rationalism, and mystics could support their otherwise antirationalist experiences with philosophical argument. Falsafah reigned for a couple of centuries in an efflorescence of arts and letters, learning and science, commerce and cosmopolitanism. (p. 231)

The term *falsafah* refers to the Islamic/Arabic passion for philosophy, and the word itself encapsulates the Arabic position—an Arabic slant on a familiar concept. Historians of philosophy debate on the significance of the Islamic philosophical achievement, with Bertrand Russell writing that not much new philosophy arose as a result of this. Russell credited the Muslims only with copying classical texts so they could later be put to use in Western philosophy. (Perhaps this is the type of argument one should expect in a book titled *A History of Western Philosophy*.)

Nonetheless, the philosopher Averroes (1126–1298), who thought and wrote in Spain, both copied Aristotle into Arabic and commented (read: explained this complicated text) on his work. St. Thomas Aquinas, the great synthesizer of the medieval system that Bacon so abhorred, relied on Averroes so much that he called Aristotle simply "the philosopher" and referred to Averroes as "the commentator."

But by then the Islamic world had turned its back on philosophy in favor of two new systems of religion and thought, both of them anti-intellectual. The conservative Islam favored by the conservative scholars called the *ulama* dominated much of the Islamic heartlands, while a hippie-ish form of Islamic spiritualism called "Sufism" that was associated with faith-healers and poets became popular and spread, especially into Indonesia, via trade.

Prior to this, in the Christian world, St. Augustine (354–430), the greatest thinker of the early medieval period, worked Plato's philosophies into Christian theology. Aristotle and his pagan materialisms won little affection from Augustine or his followers. The last philosopher to mention the great philosophers of antiquity before the late Middle Ages turned out to be Boethius (480–524).

Boethius ran afoul of a barbarian king named Theodoric, and while in prison, without the comfort of his books and apparently given little consolation by his Christian faith, he took refuge in philosophy. His book *On the Consolation of Philosophy* references primarily pagan philosophers (and the man must have had an impressive memory, not uncommon among scholars from an era where books remained rare). This mournful work focuses largely on the issue of determinism, which is not surprising considering the condemned man's fate seems so certain. Boethius lost his life, and the Christian West lost philosophy for several ensuing centuries.

Arabic translations of Aristotle went north from Cordoba, Spain, which had been under Islamic control since the Muslims drove the "barbarian" Visigoths out in the seventh century. Sylvester II, the pope of the new millennium who presided between 999 and 1002, appears to have been enamored with Arabic thought and particularly with the numbers one through nine, which featured in Islamic mathematics.

But he failed to bring Islamic philosophy to the Christian world. This task fell primarily to Gerard of Cremona, perhaps the most important translator in world history. Susan Wise Bauer (2013) writes of his contribution:

> With Gerard of Cremona, the rediscovery of Arabic texts surged forward. In Toledo, Gerard discovered a treasure trove of books he had never known existed. Among the books he unearthed in the dusty unused stacks of the Toledo libraries were a handful that had been translated from Greek into Arabic, but had never before been read in the Latin-speaking West: the *Physics* of Aristotle, containing the philosophical explorations of being that the Aristotelian texts on logic did not touch; the *Elements* of Euclid; the *Secrets* of the great Greek physician Galen. (p. 460)

Spain's history, it should be noted, was only marginally tied in with that of the Abbasid dynasty. When the founder of that dynasty, a man fittingly named Abbas, slaughtered the bloodline of the Umayyads at a dinner, one man escaped and ran to Spain. Spain, therefore, remained under the suzerainty of the Umayyads and became known in time as Al-Andalus.

Al-Andalus remains stereotyped by a general tolerance of different religious faiths (as long as Christians and Jews didn't evangelize their own faiths or criticize Islam), which was common in early polytheistic societies and uncommon in monotheistic societies, and then common again in the West after the Enlightenment. Historians also note the importance of the stunning mosque at Cordoba, and the Christian thirst to retake Spain (known as the *Reconquista*).

One can almost picture the monks of the West, educated for years on the dry writings of the New Testament and philosophically inclined toward Plato's patent absurdities, opening the dusty volumes of Arabic translations and encountering a form of thought so new to them it might as well have been of alien origin. Many of them became enamored of this new philosophy, and Jennifer Michael Hecht (2003) writes that they were in fact almost worshipful of Aristotle's work.

Soon enough, Aristotle's philosophy, rubbing up against Christian faith, caused the friction that gave birth to the fire of a quiet philosophical revolution.

This naturally occurred in France, the location of most of the medieval scriptoriums. Peter Abelard, the philosopher-cum-theologian, of the eleventh century created a workbook for his students titled *Sic et no,* or in English, *Yes or No.* This put Catholic dogma up to the test of logical analysis, and the point of the exercises was to determine whether or not the faith-based concepts of Catholicism would stand up to philosophical scrutiny.

Medieval theologians, it should be noted, tolerated a considerable amount of discussion and debate in the upper circles. As long as the discussions and potential doubts did not get to the vulgate, then nobody got hurt. Abelard took things too far. In fact, he messed with two dangerous things: the Catholic Church and a woman. The former got him banned, and the latter got him castrated.

You'll want to know more about the woman, of course, if you've not heard the story. Her name was Heloise, and she and Abelard were study partners. Despite a wide variance in their ages (he was in his thirties and she about fifteen), they must have decided to study each other because in time she became pregnant, which infuriated her family. Sources differ on the specifics, but at the behest of the family some thugs hunted down Abelard, and, well, snip snip.

Say what you want about castration, but it does nicely end a love affair. Heloise got shipped off to a convent, her baby taken from her and raised elsewhere. Abelard and Heloise continued to write letters to one another throughout their lives, and the mournful tone for a lost young love, of the kind passionate enough to shatter religious sexual guilt at a time when it seemed ever-present, maintains their ability to draw emotions from the reader.

Abelard later penned a biography, *The History of My Calamities,* which his life probably seemed like. He reminds one of the Chinese historian Sima Qian, the East Asian equivalent of Herodotus, who turned to the study and writing of history as consolation after one of the Han emperors ordered the removal of Sima's member.

At any rate, Aristotle's debut in the Christian West continued to create problems until, finally, in the thirteenth century, the Catholic Church commissioned its best mind to join Catholic dogma and Aristotelian philosophy. This legendary thinker, a philosophical hit man hired on contract to do a specific job, was Thomas Aquinas. All he was expected to do was take two totally incompatible systems of thought and meld them into a new synthesis that would become not only the definite Catholic dogma, but also the complete system for understanding the universe.

Aquinas's creation, the *Summa Theologica,* ranks as one of the most impressive intellectual accomplishments in the history of ideas. It was elegant, intricate, brilliant, and completely doomed. Aquinas himself noted the impossibility of proving faith. Faith, by definition, is a belief in something without evidence. The moment that someone has evidence for something, it ceases to require faith.

When Aquinas finished his work, the "schoolmen" that Bacon referred to, otherwise known as the Scholastics, studied the Aristotle/Aquinas synthesis and passed the teachings on without question or room for debate. Being a complete and perfect system, the notion that one could question it at all seemed heretical. This explains Bacon's distaste and disdain both for the Scholastics and for Aristotle himself. Bacon proposed a new method that restored faith in man's reasoning skills and freed the intellect from the anchor of tradition. Think *like* Aristotle, not *of* Aristotle, Bacon seemed to argue.

At this point, a definitional delineation between theology and philosophy needs to be stated. Philosophy, like good science, begins from the position of doubt. A philosopher begins from the position of not knowing anything and tries to reason to conclusions. Socrates, of course, personified philosophy by claiming to know nothing. Theology, in contrast, works in the opposite manner. Theologians already knew the ending of the argument, which was that God existed and the religious dogma represented universal truth. Theologians, therefore, employed philosophy in a defensive manner.

When doubt returned as a preposition for argument, rather than as an enemy to theology, philosophy returned to the Western world. Doubt as a concept is often said to have reappeared in the mind of René Descartes (1596–1650). Descartes worked as a mercenary, but the most violence he committed was against theology. While trapped inside during a snowstorm, Descartes pondered how he could know anything at all.

His famous proposition, phrased as *cogito ergo sum,* "I think therefore I am," written in his 1637 *Discourse on the Method,* indicated that Descartes knew that if he was thinking then he must exist. (Physicists still employ a version of this with the anthropic principle, which states that the right conditions for life must exist because life actually exists, and where the conditions are not right, no one will be there to ponder the question of life's existence.)

The doubt of Descartes found limits, however, and when he actually stared into the epistemological abyss represented by the full extent of his argument, he clutched for a handhold and found faith. Descartes wondered if it was possible that what he saw as the universe was, in fact, just a demon screwing with his head. God, Descartes declared, would not allow for such a thing to happen.

For most philosophers, Descartes eclipses Bacon. Yet it's hard to understand why, since his work has proven to be several degrees of magnitude less influential. Perhaps this is because Descartes tended to work in the more esoteric regions of thought generally claimed by graduate students in philosophy.

Bacon's occasionally vicious tone, perhaps, served a purpose. Vision and frustration are brothers. Bacon's tone may very well have been nothing more than the venting of anger that had built up over a lifetime of understanding how things should be while at the same time being acutely aware of how things actually are. What Bacon foresaw was an intellectual world animated by people actively involved in the creation of new ideas.

He understood that the only way in which his new concepts of thought would be furthered is if the practitioners received personal praise. He understood this as a kind of intellectual capitalism. Bacon despaired that those who produced ideas received very little in the way of recognition, writing: "This kind of progress is not only unrewarded with prizes and substantial benefits; it has not even had the advantage of popular applause. For it is a greater matter than the generality of men can take in, and is apt to be overwhelmed and extinguished by the gales of popular opinions. And it is nothing strange if a thing not held in honor does not prosper" (p. 70).

Perhaps, also, Bacon meant something by his occasionally belligerent tone. Too much respect for Aristotle and the scholars had cowed too many philosophers. The antidote to excessive respect, as modern New Atheists might agree were we to turn the subject to religion, may be an excess of ridicule.

Let the antidote be of a strong enough dosage to counteract the poison. Even this must be explained. Modern readers must remember that Bacon's formative years took place at the end of the sixteenth century, and that century wrought havoc on authority in general (be it philosophical, ecclesiastical, or political).

The destruction of authority in the sixteenth century might be described as starting when Martin Luther hammered his ninety-five theses onto the door of a church in Wittenberg, Germany, in October 1517 and ending when Galileo hammered nails into the coffin of the geocentric theory in March 1610. The evolution and spread of the printing press allowed for a wide dissemination of ideas, and this, in turn, created an atmosphere that gave praise, honor, and profit to people who could create and articulate ideas.

The development of the Protestant doctrine, anchored in the series of Luther's *solas* (meaning alone, as in without the Catholic Church), put an emphasis on the individual's quest for salvation and religious truth. The Protestants, in their various forms, protested the Church's overarching authority. Instead of confessing one's sins to a priest, the Protestants believed in confessing directly to God through prayer. Instead of relying on the Catholic Church's hierarchy to read the Bible and come to conclusions on doctrines, the Protestants put a heavy emphasis on literacy. Believers needed to be able to read the Bible themselves so as to reduce their reliance on the Church.

Perhaps most tellingly, the Protestant congregations chose their preachers and presbyters (elders). Contrast this bottom-up, nearly democratic function with the way in which the Catholic Church appointed priests and bishops from above. It makes sense that in the late seventeenth century British Protestants would align so closely with Parliamentarianism. The kind of people who opposed a God-ordained pope found little to like in the idea of a divinely inspired king, and the kind of people who wanted to choose their religious heads thought the idea suited politics just as nicely.

A link of antiauthoritarian rebellion can be seen between Francis Bacon's new method and Oliver Cromwell's new model army, although one can hardly imagine that the two men would have agreed on anything other than that the pope should be quiet most of the time.

Bacon lived, thought, and died in an age where the medieval authorities and institutions everywhere broke into Greek ruins; they were not demolished entirely, but the partial structures left behind no longer inspired awe so much as remembrance. Neither Christianity nor monarchy would ever look the same.

When King Charles I, slight of build, goggle-eyed, and with arrogance where his brain should have been, lost his unsightly head in 1649 to the Parliamentarians, he might as well have been standing in for the scholastic worldview. Bacon employed words, not swords, but his anger belongs to its time.

This explains the antiauthoritarian bent of Bacon's philosophy but does not frame Europe's place in the intellectual world of the time. In order to fully understand Bacon's *New Method*, a broader context of global history must be laid out.

KEY POINTS

Western philosophy begins with Socrates, whose philosophy revolved around doubt and the perpetual criticism of knowledge. His student, Plato, developed the concept of an ethereal realm of perfect forms. Plato's student, Aristotle, created a more materialistic philosophy.

The fourth-century Christian philosopher, St. Augustine, wove Plato's philosophy into Christian dogma. When Boethius died in the sixth century, the Western world's interaction with non-Platonic Greek philosophy died with him for the next several centuries.

During the medieval period, Islamic scholars preserved and expanded upon the works of the great Greek philosophers, including Aristotle. Aristotle's materialistic (in the physical sense of the word) views caused great controversy, and the Muslims eventually abandoned it in favor of more sensualist and fundamentalist forms of the faith.

Christian monks traveling into Islamic Spain (Al-Andalus) encountered Arabic translations of Aristotle. The discovery of this new type of thinking caused friction among the Catholic Church's hierarchy and her theologians. In the thirteenth century St. Thomas Aquinas melded Church dogma and Aristotelian reasoning with his masterwork *Summa Theologica.*

For centuries Aquinas's teachings were taught, by rote, to students by the Scholastics (schoolmen). Not until Galileo proved Aristotle, and thus Aquinas wrong, by experiment did the scholastic system begin to lose favor. In Bacon's time, the Scholastics still held a great deal of authority, which is why Bacon attacks them and, to some extent, the classical philosophy upon which scholasticism was based.

Ultimately, *Novum Organum* must be understood as a part of the anti-authoritarian impulse of the age. Martin Luther and the Protestants attacked the spiritual and political power of the Catholic Church, while Copernicus and, especially, Galileo attacked its intellectual and philosophical authority. Within just a few years of Galileo, Bacon recognized that the power of experimentation could be formalized into a functioning method.

THREE

History as a Connect-the-Dots Game

It is not the amount of knowledge that makes a brain. It is not even the distribution of knowledge. It is the interconnectedness.

—James Gleick, *The Information: A History, a Theory, a Flood*

Omnes scientiae sunt connexae (all sciences are connected)

—Roger Bacon, *Opus Majus*

In part II of this book, readers will be treated to an argument about the central importance of the use of analogy for teaching and understanding. There is no good reason, however, to wait until that argument is laid out to delay using an analogy for our understanding here. Understanding *Novum Organum* (I) requires that one understand the place of Europe in world history.

Understanding *Novum Organum* II requires this and more, since it is world historical forces that drive the evolution of technology. Technology drives the development of research, and perhaps equally as important, drives the creation of new analogies for human understanding.

A CONNECT-THE-DOTS HISTORY OF THE WORLD UP TO THE TIME OF *NOVUM ORGANUM*

Single-celled organisms may be thought of as dots. By themselves they are rather uncomplicated, but when they begin to combine with other cells, these dots connect into greater patterns. When multiple cells connect with one another, these too add to the complexity. Sometimes, when these organisms move into new environments, new uses are discovered for parts of the body already existing (think of how a screwdriver might be used as a weapon or a fork depending on the situation) and then those parts evolve over time to fit the new use in a more effective manner.

Other times, certain biological aspects combine with other aspects so that the combination has a new use quite different from the original use of either single aspect by itself. (Think of how wheels and engines, both created for separate purposes, can come together in a car.)

Once these organisms (dots) connect, they reach new levels of complexity. Scientists refer to the bigger picture that emerged from these connections as the Cambrian Explosion, taking place around 530 million years ago. The Cambrian Explosion represents the time period in biological history where organisms first began to connect into small pictures. Think of a really basic connect-the-dots game aimed at toddlers: The dots would be few and the connections small, and, therefore, the picture, while it would appear relatively quickly, would not be exceptionally complicated.

Watch this procedure occur in various places in various forms for millions of years, and, yada, yada, yada, we get an Australopithecine.

The historian Alfred Crosby has noted that the human ancestor in Africa, named Australopithecine, lived an arboreal life. Our ancestors likely developed the opposable thumb, depth perception, and a rotator cuff because life in the forest favored traits that allowed for swinging through trees. Crosby notes that in one of the most remarkable cases of evolutionary serendipity in the biological world, it turns out that the same equipment one uses for swinging through trees can also be used for throwing.

This ability likely explains how humans, scrawny bipeds, managed not just to survive but to flourish in a prehistoric world filled with Mastodonts, saber-toothed tigers, dinosaur-sized birds, and beavers with the equivalent mass of ecologically friendly cars. Crosby notes that the large animals in the Americas, for example, would have succumbed easily to human hunters who could commit harm from a distance.

Only smaller animals, those who could escape human hunters, such as rabbits, or deer, or squirrels, managed to survive the human onslaught. (The beaver probably shrank because humans killed the biggest, leaving the smaller ones to breed.) Humans, armed with swiveling arms, populated the planet in a geological eye-blink.

In this case, the human ability to throw is like our ancestors' ability to swing through trees. The latter equipped humans with the former, and strong evidence indicates that it was this ability to harm animals from a significant distance that allowed for early humans to migrate successfully in such a brutal prehistoric world.

Note the analogy in both the natural and technological world where an attribute that had an original purpose turns out to have another unrelated purpose. Before explaining the concepts of Jared Diamond's *Guns, Germs, and Steel* (1999), consider the fallacy in this statement, spoken by a hypothetical Californian: "People from Indiana must be really unintelligent; in the entire history of Indiana you Hoosiers have not produced a single championship surfer." The statement is immediately seen as absurd since Indiana lacks the oceans necessary for people to surf. This analogy explains Diamond's core thesis of geographic determinism.

Simply put, humans may be equal, but the geography they settled on was not. In the same way that the geography of Indiana does not lend itself to the creation of championship surfers, the geography of much of the planet was not well suited for the development of civilization. Diamond noted that the Fertile Crescent possessed the right ingredients—high-calorie wheat and the animals most naturally suited for domestication—for civilization.

Early history can be likened to a connect-the-dots game where each early civilization is viewed as a "dot," with trade and conquest drawing lines of connection between them. This did not happen everywhere, primarily only on the Eurasian supercontinent. But as these civilizations connected, it led to an increasingly complex picture, and the movement and combination of technology and ideas not only created technological complexity but made a new set of analogies available to philosophers (natural and otherwise)—but more on this in section II.

Dots of civilization formed in Mesopotamia, Persia, India, Egypt, China, and Greece, but they were mostly separate. In societies that domesticated animals (essential for a high-calorie yield and for labor) a nefarious biological equation also made itself apparent. The domesticated animals often carried disease (flu comes from pigs and smallpox from cattle) that jumped over to humans. Humans who lived in civilizations thus entered into an arms race with the viruses, a process that would go on for a few thousand years in Eurasia but that did not occur at all in the Americas, Australia, or parts of South Africa.

Although Sargon of Akkad (2334–2279 BCE) created the first empire, the first truly "great" conqueror was Alexander the Great (356–323 BCE). His soldiers carried longer-than-normal spears (an innovation Alexander carried over from his father, Phillip of Macedon) that we might think of as being like pencils. Those pencils eventually connected Greek, Egyptian, Persian, and (sort of) Indian societies.

Now, two concepts need to be defined before the narrative can resume.

The first is *cultural synthesis*. This occurs when ideas or technology from two or more cultures combine. The second is the *shoulders of giants* (SOG) principle, taken from the quote from Isaac Newton about his having seen further by standing on the shoulders of giants. In this case, the SOG principle stands for the compilation of information.

Once Alexander had connected the dots of all of the major ancient civilizations, save for China, ideas began to flow among the regions. New ideas were recorded and stored at the famous library in Alexandria, thus giving scholars access to information and creating a unique intellectual geography.

During this era, Euclid compiled his *Elements* (of geometry). Archimedes created scientific notation for the purely quixotic purpose of developing a way to count all of the grains of sand in the world. He also developed the mathematical techniques necessary to represent a sphere on a flat surface (try doing that from scratch, and you'll appreciate the problem), *and* invented water displacement theory. This last creation caused him such excitement that he leapt from the tub he was relaxing in and ran around the streets shouting "Eureka!" which is Greek for "I'm not wearing any pants!" (Just kidding—it translates to "I've found it.")

Chronologically, the Roman Republic and later the Empire come next. On the surface, the Romans present a significant challenge to the connect-the-dots thesis. After all, if the Romans connected such a massive area, then why didn't the Roman civilization produce brilliant theoreticians and mathematicians like the Hellenistic civilization did?

The answer lies with the Roman mathematical hardware. A possibly apocryphal story about Archimedes may illuminate this point. The Roman Republic invaded Syracuse in 212 BCE as a part of the Second Punic War, and according to the story, the philosopher Archimedes, whose inventions had given the brawny Roman navy fits, knelt on the ground in a temple, absorbed in an equation he had drawn in the sand.

So engrossed was he that he refused to move on the command of a Roman soldier, who ended the life of Archimedes. Plutarch gives three stories about the death of the great thinker, but only one about the grief of the Roman commander Marcellus, who was angered at the news. He knew what a rare intellect his soldiers had just erased.

The Romans brought not just conquest but a cumbersome Roman numeral system. The geographic extent of Rome may have allowed for cultural synthesis, but the compilation and understanding of knowledge was rendered impossible by the Roman numbers. Collecting modern video games does no good if one still has only an Atari to play them on.

A century and a half after the fall of the Western Empire, in 476 when the last emperor, Romulus Augustus, stepped down, a new group of connectors came storming out of Arabia, united by the newly conceived religion of Islam, and quickly conquered much of the known world. The Muslims connected the civilizations of North Africa, Spain, the Middle East, and much of Asia Minor in just a few decades. The Christian world only managed to halt the Islamic advance in southern France and at Constantinople on the Black Sea, thus preserving the area in between for "Christendom." The lands south of Christendom constituted Dar-Al-Islam, or the land of Islam.

A sect of early Muslims known to practice *falsafah* believed that abstract thinking (mathematics and philosophy) constituted the highest form of worship. Al-Mamun, the seventh caliph of the much celebrated Abbasid dynasty, had a new database constructed in the heart of the caliphate at Baghdad. This research institution/library came to be known as the House of Wisdom. Christian, Jewish, and Muslim translators copied much of the world's knowledge into Arabic and began to synthesize ideas into new wholes.

At roughly the same time, Charlemagne, the newly crowned Holy Roman emperor, sent book scouts throughout his territory for the purpose of procuring reading material for the monks in the scriptoriums. This "Carolignian Renaissance" would eventually create centers of learning that would evolve in the universities. In time, Aristotle's logic would arrive via the Muslim world, and the friction caused by doubt-based Greek logic rubbing against Catholic dogma created the philosophical fire that would eventually galvanize the scientific method.

(Al-Rashid, one of the greatest of the Abbasid caliphs, would send many celebrated gifts to Charlemagne, including a water clock. More on this in part II.)

About three hundred years prior to the creation of the House of Wisdom, the Indian intellectual geography facilitated the development of the numbers one through nine and eventually, by 500 CE, the number zero. The precepts of Hinduism encouraged the faithful to think about the long spans of time needed to attain Nirvana, and these concepts attained symbols.

The development of numbers led to a minor scientific revolution in India, and eventually, when Muslims dominated the northern section of the subcontinent, the numbers zero through nine found their way into the Islamic caliphate. Islamic scholars developed new mathematical and philosophical concepts (*al-jabr*, or algebra, being the best known among them) until stultifying religious conservatives known as the *ulama* succeeded in shutting down the short-lived era of Islamic free thought, something they continue to do today.

Up until the thirteenth century, China's civilization had developed in a separate petri dish. Rice had facilitated the development of civilization and a series of dynasties, beginning with the chariot-riding Shang, and connected the Chinese river valley cultures. By the thirteenth century the Chinese had invented not only gunpowder but also the compass and the elements of block printing as well.

Fascinatingly, the Chinese lacked the sufficient conditions to develop these technologies to their full potential. Jack Lindsay (2009) notes that since China lacked castles, there was no pressing need to develop gunpowder technology to the level where walls could be brought down. The Chinese may not have built cannons on a large scale, but they did embrace the explosive powder with all the thoughtless gusto of boys in late adolescence, blowing up live chickens, among other things. Furthermore, the massive Chinese alphabet made block printing unworkable.

In the early thirteenth century, Temujin, the greatest dot-connector of them all, took the title of Genghis Khan and led the newly united Mongols out of the steppes of East Asia. The Mongol horse soldiers, led by an evil genius, developed into the most effective military force in the history of the world. Even after Genghis died in 1227, his descendants connected (with violence that still evokes a shudder) the dots of most of the major Eurasian civilizations from Russia to China. The dots of Eurasian civilization were now connected, and Europe would prove to be the primary beneficiary of this process. Asian technology would advance in Europe, and would alter the European mind in the process.

Why Europe?

In his great book *The Measure of Reality* (1997), Alfred Crosby sums up the Western European situation during the medieval era:

> In the mid-ninth century CE Ibn Khurradadhbeh described Western Europe as a source of "eunuchs, slave girls and boys, brocade, beavers' skins, glue, sables and swords," and not much more. A century later another Muslim geographer, the great Masudi, wrote that Europeans were dull in mind and heavy in speech, and that "farther they are to the north the more stupid, gross, and brutish they are." This was what any Muslim sophisticate would have expected of Christians, particularly the "Franks," as Western Europeans were known to the Islamic world, because these people, barbarians most of them, lived at the remote Atlantic margin of Eurasia, far from the hearthlands of its high cultures. (p. 3)

Having established the backward position of the Europeans, Crosby then notes:

> Six centuries later the Franks were at least equal to, and even ahead of, the Muslims and everyone else in the world in certain kinds of mathematics and mechanical innovation. They were in the first stage of developing the science-cum-technology that would be the glory of their civilization and the edged weapon of their imperialistic expansion. How, between the ninth and sixteenth centuries, had these bumpkins managed all that? (p. 5)

Crosby explains that the civilizations of Western Europe were like stem cells, which are undeveloped but possess the potential to develop into anything. The notion serves us well here. Europe had yet to develop a civilization anywhere nearly as impressive as those of the Middle East or the Far East. But Europe, uniquely, proved to be the place best suited to deal with change, and not just changes in economics, religion, or politics, but a change in the mind. It is fair to say that the scientific method, systematized by Bacon, changed the European mind to such a great degree that many Westerners thought in a way in which Easterners would have considered alien.

The rest of the world had developed analogies they were happy with, and technology fell on stony intellectual ground there. Europe may have been undeveloped, but we can now see that she remained fertile with potential. Young people deal with change better than old people, and perhaps the same can be said of young societies as well. Francis Bacon lived, worked, thought, and wrote in a society newly changed

KEY POINTS

A connect-the-dots analogy can be employed to explain how complexity forms in living organisms and in human civilizations. For civilizations to become complex, a process of connection must occur, which then gives rise to two historical processes. Those two processes are cultural synthesis and knowledge compilation.

Cultural synthesis occurs when two societies come into contact (generally via war or trade) and technologies combine to have new uses or when a civilization evolves a new use for a technology newly arrived from elsewhere.

When knowledge can be stored, as it was in the great libraries of antiquity and the medieval era, then it can be advanced upon by each generation.

The process of connection is best facilitated on continents with a West-to-East axis, like on the supercontinent of Eurasia. The geography of Eurasia fostered the development of civilization in the same way that the geography of California might foster the development of surfers.

FOUR

Paradigm Shift

The world is stabilized, so that it cannot be moved. Who will venture to put the authority of Copernicus above that of the Holy Spirit?

—John Calvin, quoting Psalm 93:1

Ever since the beginning of the seventeenth century, almost every serious intellectual advance had to begin with an attack on some Aristotelian doctrine; in logic, this is still true at the present day. But it would have been at least as disastrous if any of his predecessors (except perhaps Democritus) had acquired equal authority.

—Bertrand Russell, *A History of Western Philosophy*

My aim is to show that the heavenly machine is not a kind of divine, live being, but a kind of clockwork (and he who believes that a clock has a soul, attributes the maker's glory to the work), insofar as nearly all the manifold motions are caused by a most simple, magnetic, and material force, just as all motions of the clock are caused by a simple weight.

—Johannes Kepler, letter to Herwart von Hohenburg, 1605

Thomas Kuhn, in his 1967 work *The Structure of Scientific Revolutions*, coined the term "paradigm" to describe the overlying scientific framework for the understanding of the universe at any given time. The most famous example of a paradigm shift took place in the seventeenth century, when Galileo invented the first true telescope and spied the moons of Jupiter.

Before moving on, it's worth pausing to note that any modern stargazer can, with only the aid of the cheapest of telescopes, gaze at Jupiter and see the three little pinpricks of light that Galileo saw for the first time. It still provides a thrill.

At any rate, Galileo's newly gathered data about the motion of Jupiter's moons, plus other data he gathered with his telescope, provided evidence enough for the heliocentric system, thus proving two long-dead men, Polish philosopher Copernicus and the less well-known Hellenic philosopher Aristarchus, to both be correct. More importantly for Galileo, he proved several other long-dead men—among them Aristotle, the Hellenic philosopher Ptolemy, and, most significantly for the time, St. Thomas Aquinas—all proponents of the geocentric system, to be incorrect.

Galileo's telescope facilitated a "paradigm shift" by intellectually scotching one model of the universe and replacing it with another. Both Kuhn and John Gribbin write of technology as being instrumental for creating these types of shifts. In the introduction to his biographical history of science, *The Scientists: A History of Science Told through the Lives of Its Greatest Inventors*, Gribbin (2002) writes:

> It is natural to describe key events in terms of the work of individuals who made a mark in science—Copernicus, Vesalius, Darwin, Wallace and the rest. But this does not mean that science has progressed as a result of the work of a string of irreplaceable geniuses possessed of a special insight into how the world works. Geniuses maybe (though not always); but irreplaceable certainly not. Scientific progress builds up step by step, and as the example of Darwin and Wallace shows, when the time is ripe, two or more individuals may take the next step independently of one another. It is the luck of the draw, or historical accident, whose name gets remembered as the discoverer of a new phenomenon. What is much more important than human genius is the development of technology, and it is no surprise that the start of the scientific revolution 'coincides' with the development of the telescope and microscope. (p. xix)

This conceit fits with Kuhn's Baconesque narrative, where technology collects new facts, which, in turn, require a paradigm shift in order to be understood. Journalists and ten-cent philosophers, and by this I mean the kind of people who scribble with markers and talk really fast on YouTube, have degraded the meaning of the word *paradigm* over the years.

It was never meant, for example, to be used in reference to a shift in economic or educational policy. The shift to the heliocentric model not only had the effect of altering the scientific worldview of the cosmos but broke the late medieval world's fascination with "the ancients" and past authorities. So diluted has the term become that a philosophical challenge to a current prevalent worldview might help to restore integrity to the word *paradigm*.

Imagine a situation where a neuroscientist/philosopher, whom we will call "Dr. Reverser," develops a theory about the human life span. According to Dr. Reverser, humans are all born in different ways and die in the same way. Each person is born with a full set of premonitions but cannot remember anything.

If someone has a scar on her arm, it is not evidence that she *has been* cut, but rather is a premonitory mark indicating that she *will be* cut. The closer that she gets to the actual "cut event," the more open and raw the cut becomes and the more vivid the premonition. Once the cut occurs, the incident is forgotten and the scar disappears.

When someone sees a pregnant woman, this is not indicative that the woman is about to give birth. Instead, the woman carries an elder in her body, someone whose life span has already played out. The elder is simply dissolving into nothingness inside the warm comfort of a carrier's womb. Once the elder dissolves, the woman will remember nothing of the ordeal, not even the pain involved from the elder crawling headfirst into the womb.

History textbooks, according to Dr. Reverser, do not represent the past but instead are evidence-based premonitions about humanity's future. We are destined, it seems, to slide into a world of lesser and lesser complexity and then to evolve into after-humanoid precestors.

If this is confusing, it's because your mind developed to think of the images in your mind solely as memories, and the notion that those images might represent the future rather than the past likely has never occurred to you or anyone that you know. Philosophy sands down all of our assumptions to what can be absolutely known. Dr. Reverser bases his theory on the fact that all that anyone ever possesses is a brief window of consciousness, something the theoretical physicist Julian Barbour calls a "now."

From the "now" perspective we have only the present and mind images. Think of a DVD put on pause: from the frozen state of the screen it is impossible to tell whether the movie is moving in forward or reverse. Dr. Reverser took something obvious and stable, looked at it from another potential point of view, and then developed a theory based on this new idea.

Dr. Reverser's hypothesis might at first seem fanciful, and too counterintuitive to be true, but what if experimentation proved his view to be supported by facts? What if Dr. Reverser's hypothesis could explain facts that the traditional view of mind images as memories could not?

The core idea ensconced in the above narrative concerns the images in the human mind. We tend to call these images "memories," and we assume, as solidly as ancient medieval people assumed the centrality of the earth in the solar system, that these images correlate with the past.

Don't despair; for reasons that will be explained in part II, there is very little chance that we human beings are all seeing our "memories" in a false manner.

However, imagine for a moment that this new conception of memories represented not just a plausible conjecture in competition with something else equally plausible but more in tune with common sense but was also backed up by hard empirical evidence. This is what the shift from the geocentric to the heliocentric worldview must have been like.

You can see now how confusing it must have been for early modern thinkers to change their paradigm from a geocentric vision to a heliocentric vision. How could the solid earth be moving when it was very clearly the sun that moved across the sky? The answer is still counterintuitive if you pause to think about it.

How many of us contemplate that the car one leaves in a parking space at work in the morning is very far away from the car that one gets into to go home at night? In relation to the trees, the other cars, and the surrounding buildings the car has not moved, but in relation to the sun, Jupiter, and other celestial objects the car has moved very far indeed. We are all like ants on an apple that has been thrown. We do not move very far in relation to the apple stem, but in relation to the objects in the room we've moved quite a distance.

Kuhn and John Gribbin alike believed that technology drove paradigm shifts. The Kuhnian conceit goes like this: Human beings conduct paradigms based upon available information. Technology makes possible the discovery of more information. New information oftentimes does not fit under the old paradigm. New information drives paradigms into a state of crisis and collapse. New paradigms are created to accommodate the new information.

Or, to use simpler language, imagine that you've connected all the dots on a page. The total picture represents a pleasant paradigm. The picture looks beautiful; it connects everything. But then (oh no!) some new dots appear on the page. If you try to connect them to the current picture you make a mess of things. You try to ignore the dots, lest they screw your picture up, but there they are, practically grinning at you. What can you do but erase the nice picture and try to create a new one that connects all the dots?

This Kuhnian view of the relationship of technology to the creation of paradigm shifts, like *Novum Organum* itself, requires not revision but a sequel for full understanding. Hold the threads of these ideas in your hand. We'll finish unspooling them shortly when we explore the concept of technology as an analogy builder.

It is up for argument as to whether *Novum Organum* constitutes a true paradigm shift. Likely not, given the definition as propagated by Kuhn, yet if Bacon's work was not a paradigm shifter it certainly altered the intellectual world in such a way that philosophers thought of themselves in a different way.

Under Kuhn's concept of a paradigm shift, technology functions as a data gatherer that produces new facts, which then require new explanations. But technology plays another, equally important role in the history of scientific thought. Technology provides analogies for philosophers to understand phenomena.

Let's borrow one of the hypothetical five-year-olds from a future chapter on the topic of Stephen Hawking and Leonard Mlodinow's theoretical position of Model Dependent Realism. Imagine we are trying to teach her photosynthesis. This is nearly impossible when she is only five because she has developed only a very few analogies based upon limited experience.

I could tell her that plants soak up the sun like she soaks up juice from a juice box (and she would marvel at how juice box analogies match up with real-life phenomena) but that would be about the extent of what she could understand. As she got older and began to understand the relationship between energy and action (such as growth), her analogies would become less naïve and more complex, and, therefore, her understanding would grow.

In this analogy, the five-year-old is like humanity, and the historical processes of cultural synthesis and the SOG principle outlined in the first section are the technologies that allow for the shedding of naïve analogies and the development of analogies of greater levels of description.

Francis Bacon famously credited gunpowder, the printing press, and the compass for remaking Europe. But did he grasp the deeper implications of this? Let's look at the impact of gunpowder.

Genghis Khan embraced a revolutionary concept of ever-changing warfare, and, therefore, he and his descendants eagerly incorporated Chinese gunpowder into their arsenal. In the year 1241, while the Mongols ravaged Hungary, Ögedei Khan died, and the Mongols retreated to solve their succession problem. In practice, they left gunpowder on the European doorstep, rang the doorbell, and ran away. Roger Bacon (1214–1294) was the first to write about this explosive new stuff, listing the ingredients as a recipe for thunder and lightning (Bacon, 1973).

Europe soon became the primary beneficiary of gunpowder. This statement might cause disruption among world historians. After all, the three Islamic empires of Asia Minor and Southeast Asia, the Ottoman, Safavid, and Moghul, all carry the historical moniker of "gunpowder empires" since their expansion was so deeply based on the use of the new technology. The reason that Western Europe benefitted the most is that the primary benefit of gunpowder was not military but intellectual.

James Burke (1985) writes that the Europeans were uniquely poised to benefit from gunpowder since they were called to church by a bell. Flip a bell over, stuff it with gunpowder, and (voila!) one has a crude version of the cannon known as a bombast. The Europeans developed the cannon to greater degrees and typically sold their technology to the sultans of the Islamic world.

A Christian named Urban, whose services had been turned away by the Byzantine emperor on account of their expense, was the one who sold Mehmet II the cannons he needed to pluck Constantinople from the branches of history in 1453.

Since Christians developed the technology, they developed the mathematical and intellectual techniques necessary to use the cannon, and this technology, in turn, turned out to have other uses. European warlords rapidly embedded the cannon into their arsenals for the purpose of shattering castle walls, thus ending the drawn-out process of siege warfare.

The effect this had on the intellectual geography of Europe cannot be exaggerated since, as Jack Lindsay (2009) has written, ballistics produced insights into motion that would affect other intellectual fields.

To begin with, cannon fire blew up the Aquinas/Aristotle synthesis, as Aristotle's notions about motion could be disproved by studying the arc of a cannonball in flight. Soldiers with cannons who use Aristotle's theories of movement would miss their targets as surely as an air force pilot would miss an enemy plane if he relied on the geocentric theory and discounted the Coriolis Effect.

More importantly, if one works out the physics necessary for describing the way that small spheres (cannonballs) move in relation to gravity, then one has paved the way for working out the physics necessary for understanding how big spheres (planets) move in relation to gravity. Gallileo began his career, we should remember, as a professor of ballistics.

Newton was a singular genius, but he did his thinking at a time when the intellectual geography of Europe had been enriched by ballistics, and Newton's elliptical planetary orbits do look suspiciously like the elongated arc of the cannonball.

Edward Gibbon (2003) may have gotten only one thing wrong in his epic work about the long collapse of Rome when he wrote of gunpowder, "If we contrast the rapid progress of this mischievous discovery with the slow and laborious advances of reason, science, and the arts of peace, a philosopher . . . will laugh or weep at the folly of mankind" (p. 1168). Science advanced not despite or around gunpowder, but because of its development.

Due to the phonetic alphabet and new techniques in metallurgy, Europeans would, in time, take advantage of block printing in a way that the Chinese could not. The compass, too, would benefit the Europeans for simple reasons of geography—it allowed the Europeans to cross the Atlantic to the new world and discover a treasure trove of historical forces. The Chinese, alas, had just been too far away from the Americas when the Ming set sail between 1405 and1433 with a set of ships that would have made Columbus blush with envy and embarrassment.

Soon enough, the inventions from the East synthesized in Europe to such a great extent that it allowed for the Europeans to set sail and guns, germs, and steel would alter the new world, while the Columbian Exchange would continue to strengthen the West. Europe, pumped with goods, inventions, and ideas from the East and West, was set to explode. Explosions, of course, do two things: they make a big mess, and they propel things forward. Europe, an unstable amalgamation, would violently connect the dots of the world.

Besides gunpowder, the clock also transformed the Western mind by creating a new set of useful analogies. According to Burke (1985), the water clock was known as the *clepsydra* (Greek for "water stealer") and probably originated in Egypt, moved to the Arabic world, and then found a home in Europe.

Europeans found it to be especially useful since cloudy days rendered the sundial inefficient. The water clock is sort of complicated to explain but really cool looking. Simply put, water fell out of a bucket or drain at a certain measurable rate and lifted a pointer attached to a buoyant ball. The pointer pointed at numbers one through twelve that had been set up beside it—clever.

There were some problems though. The clock froze in the cold European winters, and the length of the days was not uniform due to the changing of seasons. The monks needed a better clock. Also, there was a new curiosity springing up in the minds of many learned people in Europe. The monks, steeped in traditions both religious and scholarly, wanted to know more about how God's universe worked so that they could bring that sacred fire of knowledge to humanity. *What kind of watch did God have?* they seemed to be asking.

Soon enough the monks had developed a *mechanical* clock based on something called the "verge and foliot" system. This device, originally intended to inform monks of their appointed prayer hours, soon spread like the flu in a daycare. By the end of the thirteenth century, church towers featured clocks. These timekeepers had neither hands nor a face but made lots of noise to signify the time to pray.

It must have been hard to ignore the big noisy things. Astronomers noticed them and built clock faces that used the power of this new mechanism to keep the date. As Burke (1985) states:

> The earliest of these appears to have taken the form of a large face on which the pointer showed the signs of the zodiac, while windows showed other parts of the mechanism rotating the phases of the moon, the position of the sun, the major constellations as they rose and set, and certain dates, principally those of the feast days. The last was the most important function of the new clock, since the Church had a considerable number of feast days whose date depended on astronomical data. (p. 131)

Feast days were of importance not just to priests and monks but to the average worker, who found the clock was useful. What have we here? Suddenly everybody in the town could look at the big clock in the center and decide, "We'll meet at this bell at this place," which is a much more direct statement than, "We'll meet when the sun is slightly slanted in the east." Also, employers could now say, "Be here at six o'clock," rather than, "Be here at dawn." Dawn changes, but six o'clock does not.

Now time was not a fluid thing, changing with the seasons, stretching and shrinking. It had *separated* from nature and become hard, quantifiable, divided up into bits, ripped away from the universe and imprisoned inside a ringing monstrosity made up of gears and mechanical genius. Clocks rule our lives like no other invention, and any of us who have sat in a school knows their tyranny intimately.

The Europeans now had a passion for quantification. A clock is really a specialized knife. It slices time up into units called "quanta" and then measures them out in the same way that we measure space. Have you ever really looked at a clock? It's just a ruler bent into a circle, with hours instead of inches.

The clock may be the very first piece of technology to behave on the marketplace in a way that we now take for granted. That is to say that when it hit the market it was brand new, huge, and expensive. Clocks attracted tourists, but soon enough the villagers seem to have wanted the clocks they'd come to see *do* something. So many elaborations came to be added to these clocks that they bore about as much relation to timekeeping as a modern cellphone does to making calls.

As it spread it got more elaborate, cheaper, and eventually much smaller. Originally a giant machine that sat in the center of town, it would just be a few centuries before a visitor to Queen Elizabeth's court presented her, in 1572, with a bejeweled bracelet with a timepiece at the center.

How fitting that such a brilliant queen would wear the world's first watch.

In time, Isaac Newton's conception of a universe that acted like a clock would prove to be one of history's most important analogies. Religiously speaking, the deism supported by so many Enlightenment thinkers owed its existence to the clock. Deists viewed God as a clockmaker who created the clock and then stepped away from it; Thomas Paine believed that studying the clock (universe) constituted a way of worshipping the clockmaker (God).

The British theologian William Paley concluded that if one found a watch, then one may safely posit a watchmaker. Charles Darwin (who slept in the same Cambridge dorm room as Paley had while he was a student) thought differently. Darwin would have said that the clock, like all organisms, evolved. Look at the sundial, the water clock, the medieval clock, and the massive evolutionary bush of watches, digital clocks, and so on, that sprang from these origins.

Apply that concept universally, and evolutionary theory applies to everything in the world.

If the clock looks like a ruler bent into a circle, there's a good reason for that. Both concepts of measurement come from the same time period of European history where Europeans developed a passion for quantification. This brings up a question, though: If rulers measure distance, then what does a clock measure? If we assume, like St. Augustine, that time is somehow a measurement of movement, then what happens to time when there is no movement?

The current view, derived from this historical process, puts forth that before the Big Bang, nothing moved, and therefore time did not exist. This means that the Big Bang cannot be conceived of as an event on a timeline, subject to the laws of cause and effect that govern the world we live in.

This means that nothing in our cause-and-effect world provides anything analogous by which to understand the Big Bang, which is why it seems so incomprehensible. In part II, a new way of thinking about the Big Bang will be spelled out. Much becomes obvious when our current conceptions of physics are reattached to their historical beginnings.

KEY POINTS

The word "paradigm" as originally coined by Thomas Kuhn, referred to the sum total of humanity's vision of the universe. The term has since been shrunk, and thus degraded, by overuse.

Kuhn's main theme is that paradigms can shift, such as when a heliocentric conception of the universe replaced the geocentric vision. Historically, this tends to occur when new technologies (e.g., the telescope) bring in new facts via research that challenge the existing paradigm. By analogy, imagine a completed picture on a connect-the-dots page that is suddenly marred by the appearance of new dots. The picture must be erased and the dots reconnected for a new picture.

Kuhn's concept needs an addendum. Technology not only provides new research facts but new sets of analogies. The development of the clock, for example, led to the notion of a clockwork universe. It also created the concept of deism (belief in a clockmaker god) and provided an analogy for Christian theologians, such as William Paley, who believed the evidence of a clock pointed to the existence of a clockmaker and, by analogy, the existence of the universe leads one to believe in a universe maker. Darwin punctured this analogy by noting that clocks evolve, as does everything else, from simpler ancestors, and he could prove it.

The process of technological evolution can be understood by viewing world history through a connect-the-dots analogy. Technological evolution, for historical reasons, occurred with the greatest intensity in late-medieval and early-modern Europe. Alfred Crosby (1997) theorizes that Europe's lack of development aided her in the long run, since this freed her from calcifying like the ancient societies of the East. Lack of development tends to coincide with an ability to rapidly digest and benefit from change.

FIVE

Novum Organum's Impact

> The sweetest and most inoffensive path of life leads through the avenues of science and learning . . . and though these researches may appear painful and fatiguing, it is with some minds and some bodies, which being endowed with vigorous and florid health, require severe exercise, and reap a pleasure from what, to the generality of mankind, may seem burdensome and laborious . . . but to bring light from obscurity, by whatever labor, must needs be delightful and rejoicing.
>
> —David Hume, *An Enquiry Concerning Human Understanding*

LABOR OMNIA VINCIT—"WORK CONQUERS ALL THINGS"

Read Neal Stephenson's series The Baroque Cycle. This brilliant fictional series, set during the mid-seventeenth century in London during that era of Civil War (the aforementioned battles between the king and parliament), features the excitement of scientific discovery as the protagonist. Alongside this main character, readers will encounter historical characters like Christopher Wren, Edmund Halley, Robert Hooke, and Sir Isaac Newton. Francis Bacon, dead as a result of the snowy chicken incident in 1626, does not appear, but the ghost of his intellect animates both the series and the actual historical time period.

Bacon's book set the tone of the later scientific revolution, thus his extreme importance. *Novum Organum* functioned as an intellectual greenhouse, creating the right kind of conditions for ideas to grow. Philosophical horticulturalists who gaze at Newton's many-flowered *Principia Mathematica* should not forget that it grew in an atmosphere made possible by Bacon.

Here's the scene in London by the time King Charles II gave the Royal Society, the world's first scientific society, his official charter in 1660:

Twenty-one years have passed since the head of Charles I rolled away from his neck on a cold January day. Oliver Cromwell pulled the old Julius Caesar trick on Parliament, which he duly bullied into making him "lord protector." Cromwell displayed all of the personal characteristics of a seasick crocodile when he took power, beginning with his Grinch-like outlawing of Christmas (you read that right—the first "war on Christmas" came from the Puritans who saw it, correctly, as being of Pagan origin) and extending to his attempts to kill, and this is written without exaggeration, *all* of the Irish. Cromwell outlawed a lot of sinful things during his reign, and then he died.

The English people, tired of the dour Puritan ethos, essentially asked Charles I's son to just come back and reestablish the monarchy. Charles II, a party-boy fop, caused no political trouble and spent most of his time fathering bastards. Cromwell's detractors, tragically deprived of the chance to hack off his warty head during his lifetime, dug up his decomposing corpse and decapitated it. They then spiked his noggin up on a gate outside of London where it could watch the city devolve back into the sinful town it had been before Puritan rule.

Under all of this political and religious argument, somewhere beneath the decapitations and corpse-mangling, people conducted experiments and got together to discuss the implications of such experimentation. Eventually, these casual gatherings formed into the Royal Society.

There now exists a wonderful book, edited by Bill Bryson, about the history and impact of the Royal Society. Various writers of different specializations contributed essays for the book, and Philip Ball (2010) wrote one of the best of a good bunch with his "Making Stuff: From Bacon to Bakelite" in chapter 13. The essay, primarily, is about the impact of Bacon's *Novum Organum*. Ball summarizes Bacon's work:

> Bacon decries both the sterility of academic Aristotelianism, which he compares with spiders weaving tenuous philosophical webs, and the blind fumblings of uninformed practical technologies, which are like the mindless tasks of ants. True scientists, he said, should be like bees, which extract the goodness from nature and use it to make useful things. (p. 299)

Ball details how in the 1640s and 1650s, during the era of the *Interregnum* when Cromwell and the Puritans controlled British politics, Bacon's philosophy began to find receptive thinkers in the form of Samuel Hartlib, William Petty, George Starkey, and Robert Boyle (of chemistry fame). When Charles II took the throne and the puritanical Parliamentarians focused again on religion rather than regicide, these men became founding members of the Royal Society.

Ball writes that many of these early experimentalists found inspiration in Bacon's other classical work, a novel of sorts called *New Atlantis*, which, in the vein of Thomas More's *Utopia*, presented an idealized world of thinkers in the act of production as opposed to rote memorization of a system. Ball (2010) writes:

> Bacon's *New Atlantis* is a favorite hunting ground for those who like to find predictions of tomorrow's technologies. With a little imaginative license, you can find within it intimations of submarines, loudspeakers, even lasers. But even Bacon's fertile mind fails to anticipate that entirely new classes of materials might be invented. He does, however, recognize the transformative value of the textile fabrics of everyday life, and it is not hard to imagine him grasping in an instant the idea that approximations to silk might be made from oil, or the genuine article obtained without the aid of spiders and silkworms. (p. 309)

Bacon's *New Atlantis* featured a fictional and inspirational vision for the future, and one can imagine the joy that Bacon must have taken in throwing in some of his own notions about what future forms machines might take. Yet, it is a secondary work, one based upon the philosophical transformation he put forth in *Novum Organum.*

Before returning to the impact of Bacon's *Novum Organum* on the history of ideas, it is necessary to plant another seed of thought so that we can see the full-grown plant in part II. In *Novum Organum*, Bacon distinguished between the observer and the observed in a way not explored by other philosophers.

Although Descartes conjured up the fable of the sense-altering Demon, he'd not seen the full implications of his premise. Descartes still believed in God, at least in a God who would ensure that the way his senses depicted the universe would in fact correlate with the actual universe.

Bacon, who wrote before Descartes, advocated the prying apart of faith from philosophy. He believed in no such intervening God, or if he did, the notion did not infect his philosophy. Also in *Seeing Further* (2010), Rebecca Newsberger Goldstein writes of the significance of Bacon's new concept:

> The empiricist Bacon, just like the rationalist Galileo, believed that the experience that we are presented with does not reflect nature as it is: "For the mind of man is far from the nature of a clear and equal glass, wherein the beams of things should reflect according to their true incidence; nay, it is rather like an enchanted glass, full of superstition and imposture, if it be not delivered and reduced. For this purpose, let us consider the false appearances that are imposed upon us by the general nature of the mind."
>
> Bacon's solution to how to circumvent these false appearances, which he called the "idols of the cave," lay in his empirical activism. We are not to stand passively by as submissive observers of what nature might offer of itself, but assert ourselves in the gathering of facts through experiment. This assertion is what transforms sense-data, subject to illusion, into facts. (p. 8)

This is Descartes's radical doubt taken almost to its rational conclusion. Bacon wondered if the way that his senses perceive the world is in fact the way the world is. One wonders how much of Bacon's insight is owed to the shift to a heliocentric concept of the solar system. It's as if he's telling his senses "fool me once. . . ."

The problem with Bacon's concept here is that it assumes that there is a "real" way in which the universe must be, a concept that Ludwig Wittgenstein will revisit with his notion of "atomic facts." This notion presumes that the universe has an observer who perceives the universe correctly while the rest of us struggle with faulty perceptions. There is no "real" universe, only material that is then filtered through senses, but that argument will be put in place later.

For now, it's enough to say that Bacon doubted his senses and saw the cure for this doubt to be in the collection of empirical data. Data, set off by itself, would tell whether the senses lied or not. After all, it's easy to stand on the solid earth and perceive by looking at the sky that the sun revolves around us, but it was Galileo's collected data that proved his perception to be faulty. Where the senses lie, empiricism finds truth.

This is why Bacon created the concept of his "tables of discovery," which would act as a record of new empirical findings. These tables would provide a place for natural philosophers to share the results of their labors, to build upon the work of others, and to gain esteem for their works. We might think of these tables as being the ancestor of the academic journal. By doing this, Bacon had unknowingly tamed a hitherto random historical force of experimentation and discovery and codified it into a system.

Or, at least he put forth the vision. It's interesting to note that, despite all of the excitement of the seventeenth-century scientific establishment, Bacon's ideas influenced only a very small number of men. When Isaac Newton left Cambridge in 1666 to avoid a nasty burst of plague, he not only famously stuck a knitting needle under his eyeball as an experiment for his work on eyesight called *Opticks* but he also invented (along with a story of getting plopped on the noggin by an apple) calculus.

He showed it to no one. Famously, he'd later engage in a long dispute with the German philosopher Gottfried Liebniz as to who had invented calculus first, as Liebniz was having similar ideas about the same time, but we might speculate as to why Newton was originally so secretive. In fact, he probably was not; the fact was that scientific and mathematical journals simply did not exist at the time. Where, exactly, was Newton to publish his work? Who could read it, make sense of it, and share it with others?

By the time that Newton published his famous *Principia Mathematica* (with a final section titled, grandly, *The System of the World*) Londoners were tripping over coffeehouses. The coffeehouse, an idea exported from the Arabic world, provides a powerful microcosm for understanding the era in general. They only existed as a result of a new international trade network where African slaves grew sugar in Latin America.

Powdery mountains of sugar were shipped to Europe in order to sweeten the bitter drinks of coffee and tea. Meanwhile, coffeehouse patrons could also buy African chocolate, puff on a cigar made of North American tobacco, and do all of this while reading a pamphlet or newspaper made available by the development of mass literacy and the, by then, commonplace use of the printing press.

Coffeehouse owners posted stock reports. With the religious wars in decline, Europeans worshiped a new god. He went by the name of Commerce; his churches were banks, and the stock reports his gospel. The first missionaries of his word, prophets of profit, set sail about this time.

The rapid spread of coffeehouse culture can be inferred from this quote in Stephen Inwood's 2004 biography of Robert Hooke:

> London's first coffee house, Pasqua Rosée's Head, opened in St Michael's Alley, off Cornhill, in 1652, on a site later occupied by the Jamaica Wine House. Coffee houses sold exotic and increasingly popular commodities, including Arabian coffee, West Indies sugar, Virginian tobacco, Chinese tea and South American cocoa, and they offered comfortable, lively and relaxed meeting places which suited to perfection the needs of middling Londoners (rich enough to spend a few pence a night, but too poor to entertain lavishly at home). When licensing was introduced in 1663 there were eighty-two in London, and by . . . 1703 there were at least 500, perhaps many more. (p. 142)

In these coffeehouses, people talked, as they will, of all sorts of things. Apparently, they complained about the king enough to make Charles II (understandably nervous given the example of his father) issue a decree banning the coffeehouses. Londoners, good subjects, duly ignored him, and Charles let it be.

His brother James II (who took the throne after Charles died without an heir) lacked such good sense, and while James never messed with the coffeehouses, he did exert his Catholicism a little too strenuously. As a result, his own daughter and son-in-law, at the behest of a Puritan Parliament, ran him out of the country. He had to go to France of all places, while his former subjects codified their new freedoms in an English Bill of Rights. It was, in all respects, a Glorious Revolution fully in alignment with the democratic impulses of the day.

But if liberal politics advanced, science itself showed signs of stalling. It was, at this stage, still the enterprise of excited men of leisure who experimented and shared their ideas informally in the coffeehouses or at meetings of the Royal Society. Bacon had captured the historical fire of empiricism, and the Royal Society faithfully kept the flame burning, but the blaze had yet to spread. Britain would, of course, continue to produce brilliant natural philosophers, but it was the Germans who normalized the inherent philosophy of *Novum Organum* into the educational system.

In his book on the topic of Germany's tremendous outpouring of intellectual achievements from the eighteenth century onward, Peter Watson (2010) credits a change in the approach of German universities for creating Germany's intellectual dominance in the nineteenth and early twentieth centuries. He wrote that at the beginning of the eighteenth century, "the norm was the teaching of static truths, not new ideas; professors were not expected to produce new knowledge" (p. 50).

The Germans, at universities in Halle and Gottingen, invented the teacher-as-researcher concept and then brought the same approach to the students. The thinking was that teachers should be trained in the experimental method so they can create new ideas based upon empiricism.

The Germans not only turned the professor into one of Bacon's bees, but they believed that students should engage in the same process, which is where the idea of the doctoral dissertation came from. (Doctoral candidates must ask a previously unanswered question and try to fill the gap with new knowledge.)

At the same time, the academic journal was born, thus creating a historical record of ideas while also encouraging scholars to seek the prestige of peer-reviewed publication. The result of this change was what Watson called a second scientific revolution in Germany. The rest of the Western world adopted the fruitful German model, and the modern research university came into being.

At roughly the same time, people in the field of medicine began adopting and adapting Bacon's new method into medical practice. In his *The Day the Universe Changed*, James Burke (1985) writes that the influx of wounded veterans from the French Revolutionary Wars led to a revolution in French medical practice. He writes:

> In 1794 all hospitals became state property and the expansion of facilities continued. By 1807 Paris hospitals alone offered over 37,000 beds. In the whole of Britain at the same time, hospitals had room for less than 5,000 patients. The reorganization of 1794 was to make Paris the world capital of scientific medicine, attracting visitors and students from all over Europe and America.
>
> In the new *École de santé* the surgeons were now in charge. Of the twenty-two professional chairs, twelve were occupied by surgeons: anatomy and physiology, medical chemistry and pharmacy, medical physics and hygiene, pathology, medical natural history, surgery, external clinic, internal clinic and advanced clinical.
>
> The initial three-year course for students included training on tasks once thought fit for only surgeons. These included dressing wounds, making minor incisions, maintaining daily records, collecting anatomical specimens and carrying out post-mortems. The motto of the school was, "Read little, see much, do much." The success of the new ap-

> proach was immediately evident. The survival rate of fever victims being treated by physicians was much lower than those in the hands of surgeons. (pp. 206–207)

Now, this program proved to be effective but incomplete. Burke (1985) writes, "Surgeons were, after all, sensationalists by profession. Their job had always been to look, to feel, and to deal with the immediate, local cause of pain, the lesion itself." Burke writes of one of the surgeon/teachers, a man named Philippe Pinel, who in 1798 authored *Philosophical Nosology, or the Application of Analysis to Medicine*.

This widely printed and read book advocated that "concepts of sickness based on phenomena alone were inadequate. For a proper understanding of the disease the data had to be observed clinically and traced back in their sources in the organs of the body" (p. 209). In other words, a doctor's senses might deceive him, and he should gather as much evidence about a disease as possible so as to avoid falling into the trap outlined by Bacon's "idol of the cave."

Pinel may or may not have read Bacon. He may have arrived at his insights independently, or he may have been influenced by the scientific enterprise that *Novum Organum* inspired. Having looked at the impact of the science in other fields, Pinel may have thought the analogy provided would be useful for medicine.

Whatever the cause, here we have, with a rapidity brought on by the necessary work of caring for the war wounded, surgeons being turned from quacks into bees. Among the calamity of the French Revolution and Napoleonic wars, the field of medical research came to life. Any reader can recognize the medical progress made in the last two hundred years as a result of Bacon's concepts influencing the medical field.

Very few doctoral students and probably even fewer doctors realize they are working in fields that were built on Bacon's concept of the thinker as a builder of knowledge. But go into any research lab and listen. Underneath the sound of chit-chat and bubbling coffeemakers you can hear them: it might sound like the hum of computers, but don't be fooled; it's the buzzing of bees.

KEY POINTS

Novum Organum must be the most influential work in all of human history. Some version of the scientific method that Bacon inscribed in his work has been developed by nearly every society on earth, a claim that no work of religion can make.

Bacon's book, and his other works, first influenced a handful of amateur enthusiasts in London during the great era of the coffeehouse. In the eighteenth century, the Germans would work Bacon's theory of the academic-as-knowledge-producer into the university system (thus creating the modern concept of the doctorate). Slightly later, the French embraced Bacon's theories into the medical field.

These changes, gradual as they were, ripped education away from its medieval origins. Bacon's *Novum Organum* is at the center of the scientific solar system, with all subjects revolving around the concept of research as the primary way to advance in various fields.

II

The *New* New Method

> It is wrong to think that the task of physics is to find out how nature is. Physics concerns what we can say about nature.
>
> —Niels Bohr, quoted in *The Making of the Atomic Bomb*

> Thus the duty of the man who investigates the writing of the scientists, if learning is the truth of his goals, is to make himself an enemy of all that he reads, and, applying his mind to the core margins of its content, attack it from every side. He should also suspect himself as he performs his critical examination of it, so that he may avoid falling into either prejudice or leniency.
>
> —Ibn al-Haytham

> On one point not even a doubt ought to be entertained; namely, whether I desire to pull down and destroy the philosophy and arts and sciences which are at present in use. So far from that, I am most glad to see them used, cultivated, and honored. There is no reason why the arts which are now in fashion should not continue to supply matter for disputation and ornaments for discourse.
>
> —Francis Bacon, *Novum Organum*

Allow me to lay out the *new* new method in the light of a burning straw man. I do not intend here to argue for a replacement of Bacon, nor am I denying the positive impact of Bacon's schema on the world. The old new method has proved to be humanity's most beneficial idea and we should not abandon it.

Remember, I propose here only a sequel, and sequels may build upon an original without altering it. But Bacon's schema, like Aristotle's before him, is not above criticism. I certainly do not conceive of modern scientists as being anything like the dogmatic schoolmen who Bacon trounced in his philosophies.

The problem with Bacon's philosophy, and with the entire educational apparatus built on it, is that Bacon could only conceive of new knowledge as being research based. Modern universities are all built in this rut as well. When academics speak of new knowledge they conceive of it only in terms of something discovered via experimentation or research.

The limitations of such a conception became apparent to me when I began teaching. World history, the topic I teach, for example, is a research-based discipline. The educational methods courses that I have taken also fall under a research-based discipline. Historians research historical events, create an argument or narrative around them, and present those arguments and narratives in journals and books. Educational researchers study the impact of various methodologies on student learning and then publish those results. Historians and educational researchers are both bees.

But as a teacher, I was never a bee. My job did not involve researching historical documents for the creation of arguments and narratives. Nor was I researching the efficacy of teaching methods in a rigid empirical way. Instead, my job was to dive into the research done by historians, understand the concepts there, and then find ways to make this work accessible to my students using the best practices derived by educational researchers.

In other words, the work I engage in, because it produces no research information, is not considered to be a field. Teachers are not considered producers of knowledge, and this may be partially why my profession is not accorded the same measure of respect as professions where practitioners are seen also as knowledge producers. What if we shifted the definition of "new knowledge" to include a synthesis of, in this case, world historical knowledge with educational methodology? The knowledge I produce is in the bridge between the two.

By analogy, this changed concept of what it means to engage in a field of practice and to create new knowledge can be used in other areas. This concept lies in the center of the *new* new method.

Included here are complete reproductions (with a few tweaks) of two essays already published. Just to avoid confusion, Stephen Hawking and Leonard Mlodinow's book *The Grand Design* (2010) predated Douglas Hofstadter and Emmanuel Sander's work (2013) by a couple of years. Much of the *new* new method derives from the philosophy stated in these two pieces, and so I hope the reader will find comfort in knowing that these works have already been peer-reviewed and published. Their inclusion here as chapters aids the full flowering of the philosophy in the later chapters.

SIX

Everything Analogy

Some of the inventions already known are such as before they were discovered it could hardly have entered any man's head to think of; they would have been simply set aside as impossible. For in conjecturing what may be men set before them the example of what has been, and divine of the new with an imagination preoccupied and colored by the old; which way of forming opinions is very fallacious; for streams that are drawn from the springheads of nature do not always run in the old channels. If, for instance, before the invention of ordnance [cannon], a man had described the thing by its effects, and said that there was a new invention, by means of which the strongest towers and walls could be shaken and thrown down at a great distance; men would doubtless have begun to think over all the ways of multiplying the force of catapults and mechanical engines by weights and wheels and such machinery for ramming and projecting: but the notion of a fiery blast suddenly and violently expanding and exploding would hardly have entered into any man's imagination or fancy; being a thing to which nothing immediately analogous had been seen.

—Francis Bacon, *Novum Organum*

Mapping the features of one complex thing onto the features of another complex thing that you already (think you) understand is a famously powerful thinking tool, but it is so powerful that it often leads thinkers astray when their imaginations get captured by a treacherous analogy.

—Daniel C. Dennett, *Intuition Pumps and Other Tools for Thinking*

A craft arises when many thoughts that arise from experience result in one universal judgment about similar things.

—Aristotle, *Metaphysics*

To fully understand the significance of the argument made by Douglas Hofstadter and Emmanuel Sander in their new book (2013), *Surfaces and*

Essences: Analogy as the Fuel and Fire of Thinking, one must first be familiar with the fence (not a wall) separating mathematics and science from the humanities. Physicists, for example, are trained to see mathematics as the only "real" way of understanding scientific phenomena. Many great physicists explain their concepts to lay audiences using metaphors and analogies, but often regard this process as sharing Platonic shadows with those who could not comprehend the ideal. As a brief aside into the history of analogical thinking in philosophy will show, a good many serious thinkers have been widening the holes in this fence for a while now, but Hofstadter and Sander not only want to tear the fence down but also seek to have the humanities annex mathematics entirely. Central to their argument is that even Einstein's insights came to him primarily in the form of analogy, and while the authors answer no significant philosophical or theoretical questions, their argument does open up new pathways for approaching these questions.

As is often the case in philosophy, this analogical approach has intellectual ancestors. Muslim theologians used a form of philosophy called *qiyas* with analogy at its core. One could consider the fallacies developed by medieval Christian monks to be a form of reasoning by analogy, where analogies are tested to see if universal attributes can be derived.

Francis Bacon noticed this pattern and wrote about it in *Novum Organum*:

> If before the discovery of silk, anyone had said that there was a kind of thread discovered for the purposes of dress and furniture, which far surpassed the thread of linen or of wool in fineness and at the same time in strength, and also in beauty and softness; men would have begun immediately to think of some silky kind of vegetable, or of the finer hair of some animal, or of the feathers and down of birds; but of a web woven by a tiny worm . . . they would assuredly never have thought. Nay, if anyone had said anything about a worm, he would no doubt have been laughed at as dreaming of a new kind of cobwebs. (p. 79)

Bacon here intimated the importance of analogous thinking for understanding, noting that we can only understand new phenomena through what our minds have at their disposal. He later noted, while addressing the magnet and the instruments of astronomy that "these things and others like them lay for so many ages of the world concealed from men . . . altogether different in kind and as remote as possible from anything that was known before; so that no preconceived notion could possibly have led to the discovery of them" (p. 79).

Perhaps the first statement involving the concept of dual-field analogies can be attributed to the Enlightenment philosopher Joshua Reynolds, who in 1776 wrote:

> This search and study of the history of the mind, ought not to be confined to one art only. It is by the analogy that one art bears to another, that many things are ascertained, which either were but faintly seen, or, perhaps, would not have been discovered at all, if the inventor had not received the first hints from the practices of a sister art on a similar occasion. (1995, p. 349)

The counterintuitive idea here is that if a thinker wants to advance in an academic field, then she should study widely outside the field in search of analogies that can then be superimposed on problems from the original field. With this quote, Reynolds noted the centrality of analogical thinking for the sciences.

Although Ludwig Wittgenstein's famously impenetrable *Tractatus Logico-Philosophicus*, published in 1921, continues to spark controversy, largely because of its vagueness, modern forays into epistemology rescue the core of Wittgenstein's philosophy. He writes, for example:

> 2.063 The total reality is the world
> 2.1 We make to ourselves pictures of facts
> 2.11 The picture presents the facts in logical space, the existence and nonexistence of atomic facts.
> 2.12 The picture is a model of reality
> 2.13 To the objects correspond in the picture the elements of the picture.
> 2.131 The elements of the picture stand, in the picture, for the objects
> 2.14 The picture is a fact.
> 2.15 That the elements of the picture are combined with one another in a definite way, represents that the things are so combined with one another. This connexion of the elements of the picture is called its structure, and the possibility of this structure is called the form of representation of the picture.
> 2.151 The form of representation is the possibility that the things are combined with one another as are the elements of the picture.
> 2.1211 The picture is linked with reality; it reaches up to it.
> 2.1512 It is like a scale applied to reality (2009, p. 9)

In the above passage, Wittgenstein presages Leonard Mlodinow and Stephen Hawking's theoretical position of Model-Dependent Realism (MDR) by noting that the mind makes models based upon "atomic facts." Wittgenstein does not deign to explain his concepts and phrases, but by putting his words into a philosophical context, we may infer that "atomic facts" refers to the way that objects are before being perceived by human senses.

Wittgenstein's assumption, then, is that the model of reality constructed by the mind acts in accordance with the "atomic" reality. Using analogical language, he compares the mind model to reality, saying our mind models of the "atomic facts" are, in fact, *like* a scale model. Much of his argument relies on simple faith. In fact, evolutionary biology likely gives us a representative model of the universe that is grounded in "atomic" reality, simply because it is difficult to see how a radically counterintuitive mind model would be useful for helping an organism survive or mate.

In 1980 George Lakoff and Mark Johnson published the brilliant work *Metaphors We Live By*. In this book, Lakoff and Johnson explained how metaphorical thinking, such as "time is money," lies at the hearts of language and of thought. People tend to think through analogies and metaphors, and therefore the unattended mind can get caught in false metaphors. Prior to that, in 1967, Thomas Kuhn, while avoiding the language of metaphors and analogies, coined the word *paradigm* to describe the totality of the scientific understanding of the universe at any given time in his work *The Structure of Scientific Revolutions*. Kuhn theorized that paradigms often face crisis from new facts that, to borrow the language of Lakoff and Johnson, would threaten the existing set of metaphors and analogies.

The problem, from the point of view of physicists, is that the notion that analogy or metaphors might get thinkers closer to an understanding of how the universe functions than mathematics makes the scientific endeavor seem a little too much like, well, like the humanities. Hofstadter and Sander argue against this concept repeatedly in their book and are quite right to note that mathematicians and physicists are reluctant to stray from their numbers into the murky waters of epistemology or metaphysics.

A quote from the eminent physicist Leonard Susskind, in his book about his disagreement with Stephen Hawking over black holes, summarizes the attitude of most physicists. Susskind wrote that, after having finished Thomas Kuhn's classic work that the ideas of the book sparked his thinking prior to a lecture he gave in Santa Barbara in 1993:

> I would have liked to add some philosophical remarks about how evolution has created a mental picture that guides our actions when it comes to caves, tents, houses, and doors, but that it misleads when it comes to black holes and horizons. Yet those remarks would have been ignored. Physicists want facts, equations, and data—not philosophy and evolutionary pop psychology (2008, p. 255)

Susskind's disdain is an echo of Richard Feynman's famous put-down that the philosophy of science is about as useful to scientists as ornithology is to birds. Yet, Edward O. Wilson, one of the most prolific and influential scientists of the last fifty years, has recently written that mathematics really is not all that crucial for science.

In a *Wall Street Journal* piece, Wilson (2013) writes that most mathematical theorems are actually translations from already existing scientific theories. Charles Darwin, according to Wilson, bordered on mathematical illiteracy. When both Hofstadter and Wilson argue that mathematical thinking is secondary to analogical thinking, both philosophers and mathematicians should pay attention.

Furthermore, many serious physicists have started to recognize the importance of epistemology, especially when it comes to the search for the mirage of a "Theory of Everything" (TOE) that would unify classical and quantum mechanics.

In 1996, John Gribbin published *Schrodinger's Kittens and the Search for Reality: Solving the Quantum Mysteries.* That book contains a chapter where Gribbin attacks the concept of the "Theory of Everything" as a myth. Gribbin articulates his position based on Thomas Kuhn's work:

> [If] scientific knowledge really is a product of culture then scientific communities that exist in different worlds . . . would regard different natural phenomena as important, and would explain those phenomena in different theoretical ways (using different analogies). The theories from different scientific communities—the different worlds—could not be tested against one another, and would be, in philosopher's jargon, "incommensurable." (p. 198)

Here, Gribbin unites an epistemological position from the humanities with hard science, and later in his book he compares mathematics with language. (Mathematics represents a rich language indeed.) Once one accepts the analogy of mathematics and language, the next step is simply to note that languages cannot be right or wrong but more or less descriptive, and the same is true of mathematical theorems and scientific theories.

Then, in 2010, astrophysicist Marcelo Gleiser, in his book *A Tear at the Edge of Creation: A Radical New Vision for Life in an Imperfect Universe*, also critiqued the search for TOE:

> There is no final "right" to be arrived at, only a sequence of improved descriptions of the cosmos. Each era, each generation even, will describe the Universe in ways that may be radically different from the preceding one. (p. 96)

A few months after that book's publication, Stephen Hawking and Leonard Mlodinow published *The Grand Design* (2010), and echoed Gleiser and Gribbin while giving this theoretical position a title:

> According to model-dependent realism, it is pointless to ask whether a model is real, only whether it agrees with observation. If there are two models that both agree with observation . . . then one cannot say that one is more real than another. One can use whichever model is more convenient in the situation under consideration. (p. 46)

So why was the new book by Hofstadter and Sander even necessary if the concepts they wrote about seem so well developed by other thinkers? Is there anything new here? Yes. Hofstadter and Sander go further in their arguments than any previous physicist or philosopher by claiming that *all* thinking, from the superficial to the most profound, is based upon analogy.

The authors spend a lot of words leading up to their primary argument, but then devote an entire chapter, called "Analogies that Shook the World," to arguing that Einstein's insights into physics were, in fact, the elaborate creation of abstract analogies.

This argument is fascinating, because it represents a total annexation of mathematics by the humanities. Here are the authors' words:

> To some people, it might seem far-fetched to imagine that any role at all might be played by analogical thinking in the professional activity of a mathematician. After all, of all intellectual domains, mathematics is generally thought of as the one where rigor and logic reach their apogee. A mathematical paper can seem like an invincible fortress with ramparts built from sheer logic, and if it gives that impression it is not an accident, because that is how most mathematicians wish to present themselves. The standard idea is that in mathematics, there is less place for intuitions, presentiments, vague resemblances, and imprecise instincts than in any other discipline. And yet this is just a prejudice, no more valid for mathematics than for any other human activity. (p. 437)

Or, as Gribbin wrote earlier and more succinctly, "The discovery that mathematics is a good language for describing the Universe is about as significant as the discovery that English is a good language for writing plays in" (p. 199). The conceit that Hofstadter and Sander put forth here is not necessarily new for those interested in the history and philosophy of science, but the authors further the argument here to a greater degree than anyone before by claiming that Einstein thought primarily through analogy.

The authors sought to turn Einstein, typically seen as a logical and mathematical genius, instead into an exemplar for thinking analogically. For mathematicians this is not quite, perhaps, the equivalent of arguing that Jesus was a Hindu, but it's not far from it. To use but one example of the authors' argument, Einstein found an analogy between light and electromagnetic radiation. They write: "Although not in the least controversial today, Einstein's bold suggestion in 1905 that light must consist of particles was harshly and unanimously dismissed by his colleagues. Later in life, he declared this hypothesis, based upon the shakiest of analogies, to be the most daring of his entire career" (p. 460).

The rest of the argument about Einstein is deeply thought out and, ultimately convincing. It does seem, however, that the authors stopped short of spelling out the implications and applications of their thesis. In other words, for those of us who buy their argument, what does this all mean?

APPLICATIONS FOR AND PROBLEMS WITH THE ANALOGICAL APPROACH

The universe sometimes presents us with areas of time or space that defy explanation, mathematical, analogical, or otherwise; these are called singularities. Philosophy also has its singularities, such as the mystery of consciousness. The reason these singularities defy explanation for so long is that certain phenomena don't act *like* anything we come into contact with in our everyday lives. Analogies tend to evolve slowly and often in tandem with technology.

Galileo understood planetary motion only after new cannon technology helped to evolve mathematical techniques that explained how airborne spheres moved in relation to gravity. Newton developed a theory of a clockwork universe only after the clock proliferated. When Leonard Susskind theorizes about a holographic universe, he falls into the storied tradition of using technological analogies as explanatory tools.

Douglas Hofstadter has devoted much of his philosophical career to exploring the nooks and crannies of physics and philosophy while looking for connections or analogies that might be applied to deep questions. His thinking seems to be that abstract phenomena will require abstract analogies in order to be understood. For example, in his book *I Am a Strange Loop* (2007) he seemed to argue (readers of Hofstadter must sift through a lot of sand to get the gold) that consciousness can be understood as through the analogy of an information loop.

In *Surfaces and Essences*, Hofstadter and Sander argue that all thinking is analogical and that naïve analogies poison a lot of thoughts. The authors state that when schoolchildren are taught to view division as sharing rather than as measurement, it becomes harder for students to complete more complex mathematics later on. The ability to create proper analogies is therefore central to higher thought. The authors write:

> To choose one analogy over another is to favor one viewpoint over another. It amounts to looking at things from a particular angle, to taking a specific perspective on a situation. An insightful analogical take on a situation gives you confidence in your beliefs about the situation while also revealing new facts about it. A teacher, a lecturer, a lawyer, a politician, a writer, a poet, a translator, or a lover may pass hours or days in search of the most convincing analogy. (p. 382)

Before going any further, it's important to note that the authors developed a philosophy of analogical thinking that is deeper than a glib "this is like that" type that most of us associate with the concept. The philosophy, which Hofstadter and Sander argue is central to mathematics, is that analogies reach greater levels of abstraction (at some level, the ideas become difficult to convert into ordinary language). One must think about this abstraction until it becomes solid and then look to find points of connections between complex analogies in other fields. They write that, "in short, the secret of making good analogies involves making good but *more abstract* analogies—analogies between encodings, or conceptual skeletons" (p. 354).

Hofstadter and Sander note in their book that people can only make analogies based upon what they know. Individuals who have not studied deeply in a topic can only understand the topic through superficial analogies that lead to incomplete thoughts. (How many economic policies are driven by the naïve analogy that the government's budget is like a household budget?)

Bear in mind that someone could, if she wanted, make a decent explanation to a five-year-old of how photosynthesis works using boxes of apple juice. The five-year-old might think it pretty cool that apple juice boxes and photosynthesis are so closely related but then would be disappointed later by how arbitrary and incomplete this analogy is.

Hofstadter and Sander further argue that, historically, only in cases where a thinker became obsessed with a topic did it become possible for that person to see the deep analogies in that topic. When dealing with abstract questions, one might expect to find the best analogies in other areas of abstraction. This is where Hofstadter and Sander make a crucial point:

> A passion for horses or dogs does not instantly turn these animals into sources for analogies that yield insights into triangle geometry, quilt design, fly-fishing, or who knows what else. On the other hand, a *double* obsession could surely give rise to such analogies. That is, a simultaneous fanatic for, say, Euclidean geometry and for fly-fishing would doubtless find plenty of phenomena in these two domains on which to found analogies, for in this case the search would be intense on both sides. (p. 302)

And here we have returned to the concept first stated in 1776 by Joshua Reynolds. An analogical approach would require thinkers to study deeply in several fields, looking for connections between, and analogies from, each field for application to abstract problems. The future of philosophy, and the future of thought, belongs as it always has to those individuals who can think deeply in various disciplines.

This has already begun. Werner R. Loewenstein's newly released *Physics in Mind: A Quantum View of the Brain* (2013), also published by Basic Books, represents just such an attempt to use insights from quantum physics to explain consciousness. (He never mentions MDR, but does spell out the logic of the position.) Lowenstein aptly describes the actual physical makeup of the senses from the cellular level but then explains that the problem of consciousness is language very similar to the way in which Hawking has presented the mathematical issues surrounding the Big Bang. He then issues a call of a dual-field approach to the problem. Lowenstein writes:

> Regarding the nature of the very foundation of mind, consciousness, we know as much as the Romans did: nothing. We don't even have a valid paradigm. Painful as it is, I say this at the start, so as not to raise false hopes (there has been enough of that in recent years). But it would be unfair to lay all of the blame at the door of biology. The gap of knowledge, as I see it, falls as much into the field of physics as it does in that of biology. Consciousness provides a natural meeting ground for biologists and physicists—it is the ultimate frontier. (p. 215)

The problem with this notion of studying in abstract fields for abstract analogies reveals itself immediately. Can we really expect anyone to become deeply informed in both biology and quantum physics? It takes half a lifetime of study to really understand many of the concepts inherent in each field. Even if we did produce these types of thinkers, then how exactly are we to find a community of similar scholars who could peer-review the work? And where would this type of work be published?

Furthermore, Hofstadter and Sander pointed out that most analogies are false matches. Someone could spend a lifetime searching for analogies and metaphors from multiple fields but not find anything necessarily meaningful. Or, even worse, someone studying in fields far separate from his own risks becoming a dilettante and creating naïve analogies that might actually do intellectual damage. It would then require someone else to follow in the same steps across multiple fields to check or correct the theory.

One point that the authors breeze by too quickly involves the lessons that can be learned from historical geniuses. The authors state that creating a work of genius is not a deliberate process but something that comes as a natural result of an obsession with problems in multiple fields.

Historically this is true. However, there's no reason to leave this process up to serendipity. Why not capture this historical force and put it to work as an educational philosophy? The authors do not call for this, but perhaps a new approach is needed in higher education. Why not make the process of analogy-hunting a new field of knowledge in itself? Would any modern university president be bold enough to introduce a Department of Consilience and Analogy?

Hofstadter and Sander make a convincing argument, but nothing in the book indicates that they fully understood the implications of what they've done to science, or in particular, to Einstein's theories. Their intent appears to have been to show the centrality of analogical thinking even to one of history's greatest geniuses based upon the assumption that Einstein intuited great truths. Yet, the full implications of their argument would alter our perception of what Einstein did.

If Einstein thought analogically by equating mass with energy and gravity with acceleration, and, most importantly, space with time, then his set of analogies are just that—analogies that are applicable to real-world phenomena. Einstein famously conjectured that the speed of light is the universe's only constant, but he merely pulled this from his own formidable mind.

We might note that if Einstein had begun by conjecturing that the speed of sound was the universe's only constant, he could have made a descriptive theory with all of the same elements—only it would have described every particle and wave that moved *more slowly* than sound. His genius was to begin with the fastest speed that anyone knew about, maybe the fastest speed we can know about, and, therefore, created a theory that described all the particles we come into contact with.

Now, physicists assume that certain rules hold true for different levels of phenomena. To borrow the language of Hofstadter and Sander, a physicist who throws a pebble into a mud puddle can, based upon that analogy, extrapolate those same forces outward to calculate what would happen if a meteor struck the Atlantic Ocean. The same forces apply only to greater degrees. The same is true of slingshots and rocket ships.

Yet, it could be that at the particle level our analogies of propulsion, derived from the Newtonian level, do not hold up. Our confusions about the consciousness, the Big Bang, and other singularities all stem from the fact that there is nothing in our experience that provides an adequate analogy for understanding.

If this was the case, Einstein's brilliant theory would hold true for everything in the universe except for things travelling faster than what his theory forbids. If faster-than-light particles are ever conclusively found, then we will have found out that Einstein's genius was in the creation of a brilliant theoretical model, but not in the intuition of a universal truth. It could simply be that at certain speeds the analogies about propulsion that we derive from engines or muscles simply do not apply any longer.

While the ideas that Hofstadter and Sander put forth may sound abstract, in his book Loewenstein (2013) indicated that 30 percent of the U.S. gross national product is now based on quantum physics, something that just a century ago existed only in the minds of a few obscure thinkers and in the largely unopened pages of a handful of academic journals. Furthermore, the uncovering of potentially apt analogies to consciousness might yield new insights that could have benefits we cannot currently conceive of. It's not too far-fetched to imagine that someone well trained in lattice theory might be able to understand the way in which neurons are packed into the brain better than a neuroscientist who lacked such insights. An epistemological understanding that encouraged thinkers to be polymaths and analogy hunters could be as useful as anything since the development of quantum theory.

Information is now readily available, but the human mind must be able to make sense of that information for it to be coherent and useable, and this can only be done through minds that are educated to think in a certain way. Hofstadter and Sander, by annexing all fields of knowledge under the banner of analogies, have presented philosophers and scientists with a challenge that carries deep implications of change for each field of knowledge. After finishing the book, this reader is left with the impression that we still have the same old problems, but perhaps we have new way forward.

KEY POINTS

Douglas Hofstadter and Emmanual Sander's book *Surfaces and Essences: Analogy as the Fuel and Fire of Thinking* (2013) makes the case that all forms of thought take place under the banner of analogy.

This conceit echoes the sentiments of many other philosophers and theoretical physicists, but these thinkers never made the case as extensively as Hofstadter and Sander. Mathematics is a subfield to analogy, according to this new form of thought.

Even Einstein, long considered a paragon of mathematical and logical thinking, thought primarily through analogy and then translated his analogies into mathematics.

This changing conceit transforms Einstein's work into a series of arbitrary logical foundations that are then extrapolated into larger-scale theories. Light speed is not likely a universal constant but instead a brilliant place to begin a theoretical syllogism, such as, "If we take light speed to be a universal constant then we can assume," and the like.

Most importantly, polymaths can often see patterns and analogies in one field that can be superimposed to another field, thus yielding breakthroughs. This process often happens "in the wild" but has not yet been domesticated into a formal educational method.

SEVEN

Model Dependent Realism

> Many theories of the heavens may be supposed, which agree well enough with the phenomena and yet differ with each other.
>
> —Francis Bacon, *Novum Organum*

In *The Grand Design* (2010), Stephen Hawking and Leonard Mlodinow manage to give a detailed historical summary of mathematics and physics, explain the complexities of various mathematical models and theories that relate to the universe and its origin, and also find time to declare both philosophy and God to be dead. And they do all of this within just 181 pages of text. That a book so comprehensive is so comprehendible is a wonder. Yet, the presentation of the most important idea in the book, Model Dependent Realism (MDR), may leave readers thirsting for a more detailed explanation. MDR, after all, does not just provide an important perspective for analyzing physical models of the universe, it also solves most or all of the philosophical conundrums and paradoxes that have vexed deep thinkers for centuries.

MDR is really the end result of the Einsteinian Revolution. Einstein noted that when making scientific and mathematical equations, one must take into account both the observation and the observer. MDR carries this notion to its logical conclusions. To begin with, our senses evolved to make models out of sensory data in the outside universe.

Those models evolved not for the purpose of giving us a clear sense of the workings of the universe, but for evolutionary purposes, such as helping us to survive and reproduce. (Pre-Darwinian Enlightenment philosophes fretted over being limited by their senses, but lacked the insights that evolutionary biology later added.) Ancestors incapable of absorbing the light from a tightly packed group of molecules we call a rock and forming that light into a model that registers in the mind as "rock" would likely have found themselves removed from the gene pool.

All macro-organisms that survive have to find some way of detecting the sensory data around them, but the way in which those data are registered is somewhat arbitrary and likely has to do with whatever an original random mutation found useful. For example, think of a rotting carcass on a hot summer day. Clearly the carcass is giving off sensory data, but humans find that data repulsive because the meat would make us sick, but for vultures and flies the gas is the olfactory equivalent of a dinner bell. There is "material" there (the gas), but each organism makes its own "model" based on that real substance. As Hawking and Mlodinow write:

> Our sun radiates at all wavelengths, but its radiation is most intense in the wavelengths that are visible to us. It's probably no accident that the wavelengths we are able to see with the naked eye are those in which the sun radiates the most strongly: It's likely that our eyes evolved with the ability to detect electromagnetic radiation in that range precisely because that is the range of radiation most available to them. (p. 91)

Once it is understood that our "mind models" are the result of evolutionary processes much else becomes clear. Memories, for example, are just images in the mind. As Julian Barbour (2001) has pointed out, humans never really have more than just a "now" in which we live. (Defining the "now" is problematic, but for our purposes here it will suffice to say it's the shortest amount of time that a coherent thought or realization can exist. Psychologists such as Steven Pinker estimate this time to be about three seconds.)

At each "now" we can conjure up images in our mind. But how do we know that those images represent the past rather than, say, the future? What if we can see the future but cannot remember anything? What if our history books are actually books about what will happen in the future as technology gets less and less complex and archaeologists are studying the distant future? From an evolutionary perspective it is reasonable to assume that memories are connected to the past because it is hard to see how it would be evolutionarily useful for an organism to view outside "reality" in a way that was so directly contrary. In other words, if any human ancestors that saw the mind images as future premonitions were removed from the gene pool, there's probably a good reason for that.

MDR then holds that scientific and mathematical models, as well as most "inventions," are extensions of the senses. The Hubble Telescope, for example, does not actually take pictures of distant heavenly bodies. Instead, it takes in information and creates computer model pictures that are comprehensible to the humans who made it. If Hubble had been designed by heavily evolved spiders, it would have to create very different models to accommodate the sight of its many-eyed, eight-legged astronomers. The same is true for mathematical concepts and scientific models.

According to MDR, the universe may act in a certain way, but humans have to comprehend it through the mental models we have at our access. Imagine trying to teach five-year-olds about photosynthesis. You could not explain the concept to children because they would not yet have acquired enough education or experience to fully comprehend what you are talking about. As a result, you would find yourself explaining photosynthesis not as it is but in the most complex way that a child could understand it.

This may be why light is so fascinating and confusing. Even Hawking and Mlodinow seemed to have forgotten the implications of MDR when they write, "light behaves as both a particle and wave" (p. 57). Well, not really. According to MDR, light behaves like light. It is sometimes useful to describe it as a particle and sometimes useful to describe it as a wave because we lack any model that allows us to comprehend light as it really is. It's as though light is a substance that we can't stuff into two cognitive jars simultaneously, so we have to put it into one or another.

MDR can also be used to simplify Richard Feynman's theory of "sum over histories," which is notoriously hard to grasp. To return to the earlier referenced concept of time, in any given moment all humans have is a "now" of consciousness. The "now" is an island at the center (if that's the right word) of a long line of uncertainty that stretches into both the past and the future. We never actually possess the past or the future, so in its place we merely have probabilities.

To use a simple example, if I know that in a boxing match fighter A is a three-to-one favorite to beat fighter B, then all I have before the fight are odds of who might win, but not the actual winner. If I don't watch the fight live but instead record it so that I can watch it the next morning, the fact that the fight has already occurred, and therefore a reality chosen, does not affect my understanding of it. Until I watch the fight, or learn who the winner is, I am left with nothing but the three-to-one odds. Watching the fight ends the uncertainty—not for the fighters but for me.

When I watched the recorded fight, I realized that all of the possible histories (including the infinitely bizarre ones) did not occur, and the boxing match proceeded on a single course of action. Observation breaks those odds down to a single path in the same way that watching the fight makes the prefight odds meaningless. But before the fight happened the odds allowed for any possible course of action to have occurred, and I, as the observer, have to admit that anything could have happened right up to the point where I watched the fight.

Hawking and Mlodinow note that in the famous double-slit experiment in physics that "particles take every path, and they take them *simultaneously*!" (p. 75). Strictly speaking, we never know about the past of anything, so the probabilities of any possible past are infinite. Only by finding evidence can we end our own uncertainty to try and pin down a specific past.

Using MDR, we can see that it's not only light that behaves in such a "sum over histories" way, but everything else does as well. Let me return to the earlier concept of Julian Barbour's "now." If I am about to enter a mall parking lot and am undecided as to where I am going to go, then all that anyone can do is assign odds to where I might go. The odds that I would go to Starbucks would be pretty high, and the odds that I would end up on Jupiter would be exceptionally low (but not nonexistent). The closer I got to the Starbucks door, the more the odds that I would end up there increased. They would constantly recalculate until I actually entered the building, at which point the odds that I would go to Starbucks would break down and become meaningless. I had chosen a path.

The thing is, when I leave Starbucks the odds start up again. From my "now" perspective I can never be absolutely sure that I was ever at the Starbucks, because my past would have an infinite number of possibilities. Only by collecting evidence, say a credit card bill, would I be able to ascertain where I had been. Even then, all I could really say, according to MDR, is "given my memories and the evidence, I'd have to say that yesterday I chose Starbucks," but that would be subject to odds as well. Someone could have stolen my credit card, or my memories could be faulty. The odds would be against that, but it would not be impossible. Therefore, MDR shows us as conscious beings that shine lights of probability into the infinities that are both in front of and behind us

In the *Grand Design*, Hawking and Mlodinow link MDR to a version of string theory called M-Theory, because M-Theory essentially notes that all of our scientific models represent outward extrapolations of minds that had evolved to make sense of reality. M-Theory really should be called L-Theory because it treats all scientific models as languages. Languages cannot be right or wrong but more or less descriptive.

For example, Inuit might be more descriptive in a snowstorm while a Bantu language might be more helpful for describing a rainforest. Mathematical models, then, are languages that are more or less descriptive of various levels of phenomena. Hence, Newton's language is descriptive at a certain everyday level, but is not very descriptive at the quantum level. Euclid's geometry is descriptive when studying lines on flat surfaces, but one has to speak a different language when dealing with curved surfaces. The sum total of the mathematical languages and what they describe represents M-Theory.

At the end of *The Grand Design*, Hawking and Mlodinow conclude: "M-Theory is the most general supersymmetric theory of gravity. For these reasons M-Theory is the only candidate for a complete theory of the universe. If it is finite—and this has yet to be proved—it will be a model of a universe that creates itself. We must be part of this universe, because there is no consistent model" (p. 181). But I'm not sure that this is what M-Theory promises, because M-Theory is based on MDR, and as I understand their reasoning here, MDR will not allow such broad statements to be made.

Instead, MDR merely states that the universe must act in such a way that it allows for the evolution of a being that can create mental models of objects and then extrapolate those mental models into mathematical and scientific models that are increasingly descriptive. According to MDR, the universe merely has to allow for the existence of beings (and only at certain levels of reality) that can then create arbitrary levels of explanation that act as more or less descriptive languages. For five-year-olds trying to understand photosynthesis, M-Theory might include models that are descriptive from their perspective and based upon their limited experience but don't actually describe the process itself. If this is the case, we may have to confront the uneasy possibility that the universe may not make sense, even if the models we use to describe it do.

KEY POINTS

Stephen Hawking and Leonard Mlodinow (2010) created a theoretical position called Model Dependent Realism (MDR) that addresses many of the metaphysical and epistemological problems that plague traditional mathematical theories of the universe.

MDR holds that there is real physical "stuff" in the universe and living organisms, focusing on humans, creating mind-models to understand the "stuff" based upon what helps them to survive or mate.

Experience then allows for humans to extrapolate those models out into the universe. A rigorous application of metaphysics trims down the human experience to a series of frozen "nows" where we must apply statistics to apply possibilities to both the future and the past.

This notion helps us to understand that notions that were once thought to apply only to light actually can be applied to everything (thus normalizing Feynman's sum-over-histories conceit [Gleick, 1992]).

The end process of this thinking is M-Theory, which patches together all existing theories into an overriding paradigm. It should be called L-Theory because the concept treats mathematical and scientific theories as languages, which cannot be wrong but can be more or less descriptive.

By analogy, think of trying to explain photosynthesis to a five-year-old. This can only be done by explaining the phenomenon through the experiences she has at her disposal. As she gets older her analogies become more complex and, therefore, more descriptive but can never be "right."

We will never know whether or not the universe makes sense, only whether the theories we use to describe it make sense to us.

This solves many problems put forth by Kuhn. For example, the geocentric theory of the solar system explains existing phenomena (you can keep track of the stars with it) and makes accurate predictions (such as the sun will rise). So how can it be wrong? It's not; it's just less descriptive than the heliocentric system.

EIGHT

Mathematics

Any implications that logical thinking is taught only through mathematics is plainly false, and the arguments that it is taught most effectively through mathematics is, at best, questionable.

—Derek Stolp, retired mathematics teacher
and author of *Mathematics Miseducation*

What has been shown as regards mathematics as a whole through authority can now be shown likewise through reason.

—Roger Bacon, *Opus Majus*

I have left till last what is to me the strangest feature of both books [*The Selfish Gene* and *The Extended Phenotype*], because I suspect it will not seem strange to many others. It is that neither book contains a single line of mathematics, and yet I have no difficulty in following them, and as far as I can detect they contain no logical errors. Further, Dawkins has not first worked out his ideas mathematically and then converted them into prose: he apparently thinks in prose.

—John Maynard Smith (reviewing the works of Richard Dawkins)

Given the preceding arguments that human beings think primarily via analogy, that even the deepest thoughts in theoretical physics are analogous rather than mathematical, and that most mathematical theorems are in fact translations from analogy, what place does mathematics have under the *new* new method?

It would make sense in a book such as this to create an analogy. In medieval Western Europe, the original university system consisted of a regular routine, laid out here by James Burke (1985):

There were three lecture periods a day. The first ran from the morning bell at 7 am until 9 am, the second lasted from 2 until 4 pm, and the third from 4 pm to 5:30 pm. Between 9 am and 2 pm there were "spe-

> cial" lectures, or a rest period. The academic course was made up of a series of lessons, each of which took the same form: a summary of the text to be taught, the intention of the teacher regarding his interpretation of the text, the reading of the text with commentary, repetition of the text, general principles to be drawn from the text, and questions. In the evening teachers took turns at repeating the day's main points. (p. 48)

This went on for six years before a pupil could apply for an examination that would establish him as a magister (teacher). The most highly educated people in this system were called "doctors of theology." The doctors conducted the exam, and Burke writes: "The student was interrogated on the text and all relevant commentaries. The student was not asked for his own opinions, but merely to repeat what he had been taught" (p. 48). In this system, theology reigned as the "queen of the subjects" and was the last subject studied.

Theology, now, is considered an esoteric field that is typically only required in religious schools. For philosophers, theology features merely a series of bad thinking, and the fact that it is still taught as a serious discipline at all, as opposed to a historical relic, makes many wince. Imagine a class where one had to study faulty mathematical equations and you'll get the idea. Yet, at one time, those who mastered theology were considered to be the best minds in the world.

Perhaps the same is true of mathematics. If we stop thinking of Einstein, Hawking, even Newton, and so forth, as primarily mathematical geniuses and see them instead as masters of analogy, then the importance of mathematics suddenly collapses. A slew of recent articles, hardly written by anti-intellectuals, question the importance of forcing students to master the principles of algebra.

On the face of it, isn't it as absurd to force students to learn the opaque letters and numerical reasoning of a ninth-century Arabic philosopher named Al-Khwarizmi? That's what algebra, or *al-jabr*, actually is. Although these forms of math do teach, at some level, reasoning skills, those skills only vaguely match up with the kind of reasoning that allows students to analyze analogies. Even Boolean algebra is merely a set of logical symbols acting under its own internal logic.

Here's why mathematical genius is so rare. It requires people to think through analogy, find understandings, and then translate those findings into opaque mathematical jargon before anyone takes it seriously. How different is this from the medieval practice of forcing thinkers to write everything in Latin? Requiring mathematical language for the presentation of ideas and requiring students to learn algebraic equations are really no different than requiring the learning of dead Latin in an era where everybody has learned that ideas could be expressed in English just fine.

What is the primary problem with mathematics anyway? Galileo himself gives perhaps the most poetic example of a fallacy in philosophical history with this quote, written in 1623, from the *Assayer*:

> Philosophy is written in this grand book—the universe—which stands continuously open to our gaze. But the book cannot be understood unless one first learns to comprehend the language and interpret the characters in which it is written. It is written in the language of mathematics, and its characters are triangles, circles, and other geometrical figures, without which it is humanly impossible to understand a single word of it; without these one is wandering about in a dark labyrinth. (As quoted in Machamer, 1998, p. 64)

The universe is not, emphatically, written in mathematics. Neither is it written in analogy. The universe is just stuff, what we call quarks, electrons, neutrons, waves, and so on, that happens to be arranged in various patterns at various levels of complexity. We must understand it through mathematics and analogy. And the latter is a superior form of understanding to the former.

This deserves great explanation. In his book *Visions of Infinity: The Great Mathematical Problems* (2013), Ian Stewart writes that something like one hundred thousand mathematical researchers write around two million pages of new mathematics a year. Much of this mathematics has important functions, and deep mathematical thinkers can unite esoteric fields. The result of this is often new forms of technology.

Mathematicians, however, understand that algebra, for example, remains what it has always been, an arbitrary set of symbols set up in a logical or syllogistic way. Even algebra, with its bewildering symbology, embraces analogy at its core. When someone solves for X, is he not trying to find out what number X is like? And mathematicians think via cross-field analogies as well. Consider this quote from Stewart (2013):

> Historically, new mathematics often arises from discoveries in other areas. When Isaac Newton worked out his laws of motion and his law of gravity, which together describe the motion of the planets, he did not polish off the problems of understanding the solar system. On the contrary, mathematicians had to grapple with a whole new range of questions: yes, we know the laws, but what do they imply? (p. 5)

Clearly mathematics has proved to be a rich language for describing the universe, but it is incomplete, and, at its core, analogous. Besides, in the big questions, mathematics appears to be limited and a new approach (or new method) appears to be needed.

The analogy here of mathematics and Latin may prove to be incomplete. Are there deep concepts only accessible by mathematical language? Probably. But is it also possible that there are deep concepts only accessible by analogous thinking? Probably. Also, as stated earlier, mathematics is itself a form of analogy-making through the use of symbols.

In the cover essay for *Harper's* magazine titled "Wrong Answer: The Case against Algebra II," Nicholson Baker (2013) laid out the case for making algebra an elective course in high schools. Nicholson addresses the major concerns, shortened here for simplicity.

1. *Students who do well in high school algebra tend to do well in college. Therefore, required courses in algebra are a must.* Baker replies to those who push for hard mathematics standards (he calls them "standardistas") by writing: "Their endlessly repeated defense of Algebra II is derived from an obvious statistical tautology: people who take Algebra II are more likely to go to college, since Algebra II is, after all, a college requirement. In their eagerness to impose 'reasoning skills' on young people, they have in fact succumbed to an old bit of illogic: the *post hoc ergo propter hoc* fallacy of misplaced causation" (p. 36).
2. *Algebra education is essential for the nation's strength (the Cold War/Space Race argument).* Baker writes: "Think carefully again about this number: 25 percent. As the curtain rose on the baby boom era—the purported golden age of American education, when high school was really high school . . . when GM was designing the Chevy small-block V-8 engine, when missile silos held freshly minted hydrogen bombs and Admiral Hyman Rickover's nuclear-powered submarines patrolled the waves—only a quarter of high schoolers learned algebra. In the misty childhood days of IBM's Louis Gerstner . . . and of a thousand other brilliant businessmen, inventors, engineers, and innovators, algebra was a nonexistent force in the lives of the majority of high school students." (p. 38).
3. *Mathematics is essential for developing reasoning and job skills.* Baker quotes a number of mathematics teachers, professors, and practitioners who state that mathematics is not necessarily the best way to teach reasoning skills and directly quotes former Depauw professor Underwood Dudley as writing that most people will never use any math at all beyond basic arithmetic. Steven Strogatz of Cornell stated, "As someone who is working on the front lines, it's alarming to me, and discouraging, that year after year I see such a large proportion of people really not learning anything—and just suffering while they're doing it" (p. 34).

Baker suggests simply making most higher-level mathematics courses elective. This sensible solution does not degrade mathematics, but does remove it from the currently elevated seat it occupies in the American educational institutions and in intellectual life. The logic inherent in mathematics can be understood in other ways and the development and understanding of analogies can be tested in as rigorous a manner as any peer-reviewed mathematical proof.

If analogical thinking should, in most cases, take the place of training in mathematics, then how can analogies be analyzed and tested?

The answer is that students should be trained in how to see analogies, to develop them, and to determine between good and naïve analogies. This can be done by testing analogies using the medieval concept of the fallacy and the notion of consilience.

Let's begin with the latter concept, because consilience may be defined as the positive way to test an analogy. William Whewell, a nineteenth-century Cambridge don (who also coined the term "scientist" by figuring, in reaction to a criticism that natural philosophers were not philosophers, that those who practice science are like those who practice art), describes consilience as being achieved when "an Induction, obtained from one class of facts, coincides with an Induction obtained from another different class" (quoted in Snyder, 2011, p. 332).

In other words, is the concept induced from one set of facts analogous to another set of facts? This is the positive test for an analogy. If the concept is analogous across several different sets of facts, then it can be declared consilient and likely valid. Evolutionary theory, to use an illustrious example, can be induced by studying bird fossils, the bones of animals, DNA strains, or even technology or languages. Evolutionary theory is so consilient that it could be argued that every single thing that one encounters abides by its logic.

That last statement might sound like a stretch, but nothing you will ever encounter lacks an ancestor. The cellphones we talk on have myriad ancestors dating back at the very least to the first antiquated phones found hanging on the wall in Andy Griffith reruns. An evolutionary bush of different phones evolved from this common ancestor. The camera function on that cellphone can be traced to the first cave paintings, through the flat medieval pictures, to the Renaissance concept of the vanishing point, and then up to the first daguerreotypes and early forms of flash photography.

Edward O. Wilson (1998) fully explained and then expanded the concept of consilience in his book *Consilience: The Unity of Knowledge,* which features as a polymath's come-to-Jesus. Those who study across fields find unique points of contact, according to Wilson. The way in which to judge analogies is to test them for consilience across disciplines.

Nothing revolutionary is inherent in this concept. Think of a situation where a man is on trial for having stolen a loaf of bread in order to feed his hungry son. The prosecutor will try to state that this case is analogous to a previous case (a precedent) where a mother stole a candy bar to prevent her daughter from going into insulin shock and was found guilty. The defense attorney will argue that the case is more analogous to a situation where a grandmother stole a bottle of water to prevent a toddler from dehydrating on a hot summer day and was found innocent.

The question here is whether or not the concept of guilt or innocence involving theft in cases where the thief has an altruistic intent in an immediate case is consilient across like situations.

Allow two quick examples here before the more scholarly presentation. A recent documentary about steroid use and abuse among athletes was rife with false analogies and bad reasoning. The show's producer interviewed classical musicians about whether or not they took antianxiety pills before performing.

Most did and saw nothing wrong with this. Well, went the producer's reasoning, if it's okay to use performance-enhancing antianxiety drugs for musicians, then why can't performance-enhancing steroids be used by bodybuilders and power lifters?

The problem here, of course, is that the effect of antianxiety drugs on musicians and the effects of steroids on bodybuilders are not analogous. One should always reduce factors when comparing analogies. So the direct analogy would be to compare the effects of antianxiety medication for musicians with the effect of antianxiety medication for bodybuilders. This analogy brings us closer to minimizing factors that might skew the analogy.

The use of analogies is central to ethics. Many a fight amongst young couples in love begins with the phrase, "How would you like it if I ____." So the boyfriend may stay out partying all night, and then, when he stumbles in, his irate other half might ask how he would like it if she stayed out partying all night as well engaging in other analogous activities.

Consilience is the positive way to address analogies, and if concepts hold true across a variety of facts and are, therefore, analogous, then they can be declared valid until reason is given to no longer believe this.

THEOLOGICAL PHYSICS

The term *theological physics*, almost certainly put in use here for the first time, should not be interpreted to mean that physics and theology, or science and religion, are in any way compatible. The term is used here, somewhat playfully, as a way of noting that concepts from medieval theology and philosophy can be used to determine the usefulness of an analogy. The rules developed by medieval theologians can be understood as negative ways of testing analogies.

Theologians once sat atop the medieval university system's hierarchy, and intellectuals considered them to be the deepest of thinkers. In the late thirteenth and early fourteenth centuries, Aristotle's logic caused a break between the proponents of an older theological system (those who tried to reconcile faith and reason) and the primarily secular Aristotelian philosophers, among whom William of Ockham remains the most notable and influential.

Modern philosophers, unimpressed by Aquinas's watery logical proofs for the existence of a god and generally suspicious of a thought system that ostensibly placed faith above reason, tend to think poorly of theology if they think of it at all. While some of the derision is merited (especially that aimed at John Duns Scotus, whose impenetrable writings read like the wiki posts of a doddering English academic after he's drunk a quart of schnapps and suffered a mild head trauma), much of value can still be found in the dense writings of medieval philosophy.

Mathematicians and historians of science have long puzzled over the nature of infinity. How can it be that real whole numbers like 1, 2, 3, and so on, can go on to infinity while the decimal points between those numbers can, as well, be divided infinitely? How can we have infinity inside a finite space? Georg Cantor, a Russian-born German, posited in 1874 that infinity comes in different sizes. Cantor's name remains a fixture in the mathematical hall of heroes.

Yet, here is the medieval philosopher Robert Grosseteste (1168–1253) writing:

> But it is possible that an infinite sum of number be related to an infinite sum (of number) in every numeric ratio and also in every non-numeric ratio. And there are infinities which are greater (plura) than other infinities, and infinities which are smaller (pauciora) than other infinities. For the sum of all numbers both even and odd is infinite. For (the sum of the even and odd) exceeds the sum of the even by the sum of all the odd numbers. Moreover, the sum of the numbers double continuously from unity is infinite; and similarly the sum of all the halves corresponding to these doubles is infinite. And the sum of these halves is necessarily half the sum of their doubles. Similarly, the sum of all the numbers tripled from the unity is three times the sum of all the thirds corresponding to the triples. And the same thing is clear in all species of numeric ratio, since the infinite can be proportioned to the infinite in any of these ratios. (pp. 475–476)

Clearly, medieval philosophers had interests beyond the creation of theological proofs, and their thoughts frequently wandered into the secular and profound. Importantly, for the *new* new method, medieval philosophers evolved a way to test analogies by using negation, or fallacy, as a parameter.

In order to explain, our history of Aristotle must be revisited. When Aristotle, newly arrived to medieval universities via Arabic translations from Cordoba, broke open theology, as James Burke (1985) writes of the consequences:

> The arts faculty soon became controversial because it was here that the full effect of the new knowledge from Spain was felt most strongly. Students were trained to examine nature textually in the *trivium*, and

> through the use of mathematics and reason in the *quadrivium*. Also taught was logic, which, thanks to Aristotle, was fast becoming the most revolutionary of subjects.
>
> While the new learning stimulated the creation of universities, it posed fundamental problems for the Church. (p. 49)

The key problem from the Church's point of view was exactly what was represented in Peter Abelard's *Sic et no* (*Yes and No*). One could examine doctrine and find it either in alignment with logic or not. This was a process originally understood by the Arab Muslims who had an entire school of reasoning which made use of analogy.

Think of *Yes or No* as a form of judging analogies. If a form of reasoning fits in this situation, does it also apply to another similar situation? In other words, does a form of reasoning hold true across a variety of analogies? If it does, then it is reasonable. If not, then no.

You see, when the memorization of texts and dogmas reigned supreme as an educational theory, then theology stood atop the intellectual mountain.

When an emphasis on Aristotelian logic changed the emphasis in education, then syllogistic reasoning became the basis of education, and mathematics rose to the top. Now that analogical thinking is understood to be, as Douglas Hofstadter and Emmanuel Sander (2013) put it, "The Fuel and Fire of Thinking," then it's time to replace mathematical thinking with thinking by analogy.

And if analogy is to sit on the throne, then we must develop ways of telling whether or not the king wears clothes.

False analogies plague thinking in ways that make a logician want to cry. (Read *Crimes against Logic: Exposing the Bogus Arguments of Politicians, Priests, Journalists, and Other Serial Offenders*, by Jamie Whyte, to see a bunch of public examples.) Abelard gleaned the concept with his *Yes or No*. What he was asking then is what we should be asking now—is the concept to be studied analogous with other facts?

About this time, Roger Bacon (1214–1292), who is of no known relation to Francis, argued for the primacy of mathematics in Western education. He wrote: "Wherefore it is evident that if in the other sciences we should arrive at certainty without doubt and truth without error, it behooves us to place the foundations of knowledge in mathematics, in so far as disposed through it we are able to reach certainty in other sciences and truth by the exclusion of error. This reasoning can be made clearer by comparison" (1973, p. 483).

What Bacon meant here is that the truths of mathematics could be tested against other truths to see if the principles discovered there were in fact analogous to one another.

Medieval reasoning, where false analogies are labeled as absurd, could aid in this process of analogy judging, as could the concept of consilience. The most famous Jewish philosopher of the medieval era, Moses Maimonides (1138–1204), worked almost entirely through the use of negative analogy.

Maimonides reasoned, eventually, that God could not be understood or even properly discussed because nothing in our experience prepares us for the analogy. (God as a singularity!) This passage from the great philosopher's most famous work, *A Guide to the Perplexed*, is worth quoting at length:

> It is a mistake to attempt to prove the nature of a thing in potential existence by its properties when actually existing. If you do this you will fall into great confusion, rejecting evident truths and accepting false opinions.
>
> Suppose a woman gives birth to a healthy baby boy but she dies after nursing him for only a few months. The father raises the boy on a lonely island where there are no other females. The boy grows to be a wise man, who nevertheless has no idea how human beings come into existence. He asks someone and is told that human beings begin their existence as very small living beings in the womb of a woman, where they move, receive nourishment, and grow, until they finally come out. The man will naturally ask whether these small beings breathe with their mouths and their nostrils while in the womb. The answer will be no. So then the man will undoubtedly attempt to refute the explanation and prove its impossibility by referring to the properties of a fully developed human being. He might say 'When any one of us is deprived of breath for a short time we die and cannot move any longer. How then can we imagine that any one of us has been enclosed in a bag in someone's body for several months while remaining alive and able to move? If any one of us would swallow a living bird, it would probably die immediately when it reached the stomach, or at least when it came to the lower part of the belly. This mode of reasoning would lead to the conclusion that human beings cannot come into existence and develop in the manner described.
>
> If philosophers would consider this example well and reflect on it, they would find that it represents exactly the dispute at hand. (As quoted in Kaye, 2008, pp. 108–109)

This dense extract includes analogies within analogies. The isolated boy cannot believe that humans come from a woman's womb because nothing in his background provides an analogy. He searches for one involving an unfortunately swallowed bird and, therefore, reasons to the wrong conclusion. He does not, and cannot, know how naïve his bird-swallowing analogy is.

He then exhorts the reader to view the entire dilemma as being analogous to the philosophers' attempts to create a logical proof for god. Nothing in human experience can be analogous to an Almighty power, so the endeavor must be called off or at least understood for what it is.

To restate an earlier point, modern philosophers and physicists struggle with the same concept. Except, instead of trying to find analogies for a god, modern thinkers seek to find analogies for singularities.

Before beginning on a formal statement of analogy it might prove useful to note that, at a lesser level, humans test analogies all of the time. Legal reasoning represents a shallow form of analogy testing in that lawyers try to poke holes in each other's analogies. Most are not very good at it.

A modern school of constitutional law, for example, holds that the Constitution is a dead document to be interpreted as binding only in relation to societal norms in 1789. The Eighth Amendment's guarantee that American citizens should be protected from "cruel and unusual" punishment only applies, according to this school of thought, to what was considered to be "cruel and unusual" in 1789. If the times change then lawmaking bodies should be free to pass laws that would redefine that.

Yet, judges who opine under the banner of this conservative ideology refuse to allow for any laws that would restrict the Second Amendment's "right to bear arms." The contradiction becomes immediately apparent. If the Eighth Amendment applies only to 1789's norms, then the logic should apply to an analogous situation as well.

The Second Amendment should be read as applying only to the people's right to bear muskets or any other military technology available to citizens at the founding. The Constitution would not guarantee anyone's right to possess even a deer rifle (though laws could be passed allowing their possession).

The logic applied for the Eighth Amendment is not analogous to the Second Amendment and therefore fails to maintain its internal structure. The legal reasoning itself may or may not be valid, but if even the proponents cannot apply its logic across a variety of situations, then the logic falls into contradiction. Those who argue that atheists should be unbothered by hearing the phrase, "under God," in the Pledge of Allegiance should consider an analogous situation where religious children would hear a pledge that included the phrase, "one nation which denies God exists," before every school day.

Consider another legal/ethical argument, all too often pronounced by religious conservatives concerning homosexuality. Those who condemn homosexuality often link it with bestiality and incest. These are all sexual crimes condemned in the Old Testament. One cranky Supreme Court justice has even lumped homosexuality together with murder or theft, with the intent to categorize homosexual behavior as something that can be freely condemned by society.

The problem is that homosexual behavior lacks a victim and, therefore, is not analogous to bestiality or incest. Natural rights theory holds that individuals each possess rights that cannot be infringed on by others.

Lawyers sometimes state this simple principle with the dictum, "Your right to swing your fist ends where my nose begins." One has a right to exercise freedom only up to the point where it interferes with someone else's freedom.

Homosexuals violate no one's freedom and cause no victimhood, and, therefore, homosexuality is not analogous to other crimes that include victims. Those who claim that homosexuality violates God's law as put forth in the Old Testament would then have to explain why *all* of God's laws (including the execution of children for talking back to their parents) should not be woven into American law.

Medieval philosophers practicing the process of testing analogies noticed that the same mistakes kept appearing and named those errors. These logical tests provide much of value for testing modern analogies.

Maimonides and William of Ockham (1280–1349) provide the most useful of negative tools for testing analogies. Maimonides, because he believed nothing in human experience could be analogous to God, focused on describing what his God could not be. Sherlock Holmes later famously lectured his sidekick Watson with the phrase, "When you have eliminated the impossible, whatever remains, however improbable, must be the truth." This sums up the approach of Maimonides.

The problem, as William of Ockham and other philosophers have discovered, is that once the sharp edge of logic begins its work on dogma, the impossible remainder is nothing at all. Or, to quote John Buridan (1300–1358), "I say that god is most of all simple, since He is not composed of parts, nor can He be composed of anything" (1973, p. 762). Buridan surely did not consider himself an atheist, but he understood that if he claimed that God was composed of matter, then he would strangle his deity with contradictions. Even Aquinas noted that the task of describing what a god isn't is considerably easier than describing what a god is.

William of Ockham did not create the philosophical razor he is most known for but did expound its basic philosophy most eloquently. Ockham helped to shape a "new method" of sorts himself, being a member of what philosophers call the Nominalist or modern school of philosophers who distinguished themselves from the old-school theologians. The mark of this school of thought involved a willingness to follow the precepts of reason further than previous theologians had.

He did state, however, that "plurality is not to be posited without necessity" (1973d), but this merely gave linguistic form to a long established principle. His slight-looking razor slices away the supernatural from reason. One cannot posit invisible entities beyond what the evidence calls for. Doing this for explanatory purposes causes a "plurality" or excess, of explanations not needed. The nut falls from the tree because of gravity, not because of gravity and because Thor wanted a squirrel to have the nut and whacked the tree with an invisible hammer.

Ockham's greater significance in medieval philosophy really stems more from the fact that he fell most definitively into the secular camp of logical theologians. Ockham carried a razor, but it took the form of his unique and historic intellect. The word "god" is a rare spice in his writings. The modern reader wonders how much of a believer William actually was since much of his writing undermines the Platonic ideal upon which most of medieval theology teetered. Here he is slicing up Plato's concept of the forms:

> Creation is absolutely from nothing, so that nothing essential or intrinsic to the thing precedes it in real being. Therefore, no non-varied thing pre-existing in any individual belongs to the essence of this freshly created individual, since if anything essential to this thing preceded it, then it would not be created. (1973a, p. 663)

See the metallic flash of the razor just before he plunges it into the heart of the Trinity:

> Concerning this question it should first be understood that for the intellect it is not a question as to what should be maintained according to the truth, but as to what one would maintain who wishes only to be supported by the reason possible in this present condition, and who does not wish to accept any doctrine or authority, just as who did not wish to accept any authority whatsoever would say that it is impossible for three persons distinct in reality to be one supremely simple thing. (1973a, p. 679)

This philosopher's hand is red with the blood of Christian dogma. In the above, Ockham asserts that one could not reason to the Trinity and that, unless one simply believed it because he was told to, then no good logical reason for belief could be asserted.

His enemy bleeding, Ockham neatly slices both the jugular of his subject and the link between faith and reason with this stroke:

> I reply otherwise to the question, that whatever the truth may be, one wishing to be supported by reason, so far as it is possible for a man to judge from what is purely natural, in this present condition, would more easily deny that any relation of the genus of relation is a different thing, as previously expounded, than maintain the opposite. And this is because the more difficult arguments are for this side rather than the other. Indeed, I also say that the arguments for proving such a thing which are not supported by Scripture and the sayings of the saints are in no case fundamentally efficacious. And hence I say that just as one who would wish to follow reason alone and not accept the authority of Sacred Scripture would say that in God there cannot be three persons with a unity of nature, so one would wish it to be supported only by the reason possible for us in this present condition would equally have to hold that a relation is not any such thing in reality as many imagine. For no inconvenience for the negative side follows from principles known from what is purely natural and not taken on faith. Nor can it

> be shown through reason that not every thing really distinct from another is thus an absolute thing. (1973a, p. 681)

One cannot reason to precepts of dogma. You cannot have your faith and prove it too.

A complete list of the logical fallacies would exhaust the reader here and prove ultimately to be unnecessary. One does not need to master all of the logical fallacies in order to develop the form of thinking necessary to engage in the *new* new method, but a short list of the most common fallacies should suffice to give over the general idea. The point of listing the fallacies is to restate them in the phraseology of the *new* new method so that these fallacies can now be understood as means for testing analogies.

AD HOMINEM (TO ATTACK THE MAN)

This fallacy occurs when one fails to address the argument or analogy put forth by another person and instead tries to discredit the person making the argument. An example of this is if a philosopher expounds a position and then her opponent states, "This is the kind of argument we would expect from someone who had three demerits on his permanent record from grade school."

The argument is not analogous to other situations. Could we, for example, attack relativity theory by saying of Einstein, "This is what we might expect from someone who styles his hair by licking a light socket." The personal attack on Einstein's hairstyle would have no effect on his theories. Therefore, the logic does not work in analogous situations.

APPEAL TO MYSTERY

Whenever a scientific explanation is not available, many people look to mysticism for an answer. The best example of this fallacy relates to the Big Bang. The logic goes like this: Science cannot explain thoroughly how something came from nothing; therefore, a supernatural deity must have created the universe.

This line of reasoning is not analogous to other mysteries and is, thus, invalid. Trying to use the appeal to mystery on more mundane subjects reveals its absurdity. For example, imagine a situation where you cannot find the remote control. The remote's location is, therefore, a mystery. One would not find it very useful to assume that something supernatural happened to the remote. The logic for a big mystery cannot be used on analogous situations involving small mysteries.

(And yes, those little cuts are from Ockham's razor, which would slice away excessive pluralities in the explanation. One will never find the remote if one posits, without evidence, that leprechauns took it to Ireland.)

REIFICATION

Plato's fallacies all grow from the same tainted soil of reification. This modern term, not created during the medieval philosophical period, involves the error of believing an abstract concept to be a real thing. For example, when a reporter states, "The economy was sluggish today," she reifies an abstract concept. The "economy" is not a material thing to have emotions or physical states. It is not like a real body. The "economy" merely refers to a series of individual monetary interactions among individuals.

This may seem a niggling point, but the effects can be pernicious if we are not careful to purge ourselves of reification. To say, for example, that "the Fed raised interest rates" assumes an image that the Fed is a faceless thing, rather than a compilation of (wrinkly) individual faces. The effect on thinking can be serious.

Modern philosophers who posit a multiverse commit the sin of reification most seriously. The thinking is that the "laws" of our universe seem closely calibrated and therefore require explanation by way of increasing the odds. To use a simplistic analogy, if you hit the lottery you'd know that a bunch of other people did not. The odds are that someone had to hit the lottery, but your hitting it as opposed to someone else could be explained by invoking the anthropic principle.

Yet, the multiverse analogy needs work. We hold the winning lotto ticket (a universe with calibrated laws that allows our existence) but have no physical evidence for any other universe(s). We merely assume that something unlikely must have a multitude of other unlikely configurations, in the same way that a winning lotto ticket is just one configuration among a multitude of losing tickets. We have no evidence for any other universe and, therefore, do not know how unlikely our calibration is.

The analogy for trying to assign odds where we have information about all possible types of the way things could not be (e.g., the losing lotto tickets) to a situation where we have evidence only for the way things can be (e.g., our universe) does not apply, and, therefore, the reasoning is false.

CIRCULAR REASONING

All truth flows from the _________ (Bible, Koran, works of L. Ron Hubbard) because the _________ (Bible, Koran, works of L. Ron Hubbard) has claim to the sole fount of all truth. The fallacy also applies to talk radio. This may be why religious wars used to be so common. Circular reasoning can survive if no other religions exist to make a mockery of the analogy. "Catholic" was originally an adjective meaning "universal," and only after the Reformation did it become a proper noun.

If the Catholic Church remained the only authority, then the idea that "the Church hierarchy is correct because the Church hierarchy is infallible and the Church hierarchy says it is correct" makes a certain sense. It's only when competing churches make the same claim that one realizes that the logic for one church is not analogous when applied to another church.

They can't both be infallible and say different things.

FALSE CAUSATION

This occurs when someone asserts a cause that has no reasonable connection with the effect. Football fans that attribute the wearing of a favorite sweatshirt to their team's ability to win or lose games fall for this fallacy. Would the favorite sweatshirt work in an analogous situation?

The fallacy is sometimes harder to see. We tend to assert that good outcomes must come from good causes. Society is less violent than it once was. Is this because humans have become more enlightened, or is it because young men are now too chubby and lobotomized by computer games to bother with criminal acts?

A list of the fallacies quickly becomes tedious because the alert reader can quickly get the gist of the process. Is something true simply because an authority figure said it to be true? Can we think of an analogous situation where an authority said something that turned out to be wrong? Of course. So this is the fallacy of authority. Is something true just because the majority believe it to be true or because someone threatens force if we do not believe it?

The fallacies serve merely as a starting point for those thinking within the parameters of the *new* new method. Serious thinking at the heart of analogies requires a more intimate knowledge with specifics, and this takes the philosopher into deeper territory.

For example, much controversy flows around the concept of the "ten-thousand-hour rule," which states that mastery of any activity usually requires up to ten thousand hours. Meant as a rough estimate, the concept has been ridiculed by those who point out that some high jumpers or sprinters can compete at a high level with virtually no training. Certainly, a seven-foot man could compete at a high level of basketball with much less training than someone of average height.

But is athletic performance analogous with piano playing, deep theoretical mathematics, or novel writing? We see, for example, talented child actors and mathematicians, but never talented child novelists or playwrights. Could it be that some activities (like jumping, running, and acting) are "easier" since they are more closely related to what our genetics prepare us for and, therefore, more affected by nature than nurture, while other activities (like writing and philosophizing) are quite far away from what our genes intended and therefore more likely to require a more uniform number of practice hours for most people?

Are certain activities more likely attuned with the ten-thousand-hour rule than others? Is the rule more applicable the further removed the activity is from anything our ancestors would have engaged in and, therefore, removed from the natural shaping that evolution provides? The thought process involves the creation of principles on one set of facts and then testing that principle as analogous to other sets of facts.

The more we know, the more intricate the analogies. Cancer research already operates under this idea. When someone is diagnosed, the question is, "How like other cancers is my cancer, and how successful have the treatments of those other cancers been?" With only rudimentary knowledge, a tumor might be lumped in with all kinds of other tumors that it does not really belong with. As we know more about the molecular and genetic structures of tumors, we can categorize them more effectively and construct better analogies.

As it stands, the treatment for a certain type of tumor might not work on another because the analogy is naïve. Creating more effective analogies will help to treat individualized tumors as long as a sample size exists that is sufficient for the creation of analogies. This example reiterates that the *Novum Organum II* is intended as a sequel and that the collection of data, which is central to *Novum Organum* (I), remains essential.

To give some idea of how a medical practitioner of the *new* new method might incorporate the new learning with the old, let me borrow an example from a modern medical classic, Anne Fadiman's *The Spirit Catches You and You Fall Down: A Hmong Child, Her American Doctors, and the Collision of Two Cultures* (1997). The book features the story of Lia Lee, a child of the immigrant Hmong peoples from Laos. Seizures attack Lia frequently as a baby and toddler, shaking her with frightening violence.

Her parents, raised in a pocket of the world isolated from the impact of *Novum Organum*, do not follow the medical advice given by Lia's well-meaning doctors and fail to give Lia her anti-seizure medications. Lia's main doctor reports this to child welfare services, and the government takes Lia, then a toddler, from her parents, giving her to a foster family for a time.

Under the *new* new method we can see that when Lia's doctor encountered parents who would not give their daughter the right dosage of medication, he reached into his analogical experience for similar occasions. In his experience, parents who would not give their children medicine were negligent at best and abusive at worst. He applied this analogy to Lia's parents, who, as Fadiman tells the story, may have been confused but were anything but negligent or abusive.

What if Lia's doctor had been able to see the impact of this bad analogy on his decision making and then on Lia and her family? Even with medication, a seizure of Richter scale proportions eventually destroys her cognition and personality, rendering her "brain dead." This may have been unavoidable no matter what, but she and her parents may have been spared pain if a different operant analogy had been put in place. How many ethical decisions can be altered by applying a new form of metacognitive thinking?

There's a lot that can be done with the *new* new method.

KEY POINTS

Mathematics enjoys its place at the top of the intellectual pyramid largely because few people understand deep mathematical reasoning (thus mystifying it) and because of the high place that logical rigor holds in the sciences.

Many mathematicians believe that math is specially aligned with the universe, which is not true. Mathematics is an arbitrary (but descriptive) language that can be applied to phenomena. The same work can be completed using analogies.

Analogies can be tested as rigorously as mathematical proofs by using the test of consilience and medieval forms of philosophy. Can the analogy be applied across a wide assortment of facts? Does the analogy fall into any of the categories defined as fallacies? These questions provide analogous thinking with rigor.

Mathematics is important for collegiate achievement because math is, inexplicably, considered important by college entrance committees. Most mathematics classes could be replaced with analogy-testing courses, which would teach logical reasoning more efficiently.

Math should not and will not go away, but under the *new* new method it will have to be removed from its throne.

Analogy-making is not just useful in the scientific or academic sense but can help open the mind to "meta" types of analysis. Understanding the difference between good and bad analogies helps us to make better decisions, ethical or otherwise. Training in this type of thinking is much more useful than mandatory classes in algebra.

NINE

Examples from the Field of Analogy and Consilience

> Such is the infelicity and unhappy disposition of the human mind in this course of invention, that it first distrusts and then despises itself: first will not believe that any such thing can be found out; and when it is found out, cannot understand how the world should have missed it for so long. And this very thing may be justly taken as an argument of hope; namely, that there is a great mass of inventions still remaining, which not only by means of operations that are yet to be discovered, but also through the transferring, comparing, and applying of those already known, by the help of that learned experience of which I spoke, may be deduced and brought to light.
>
> —Francis Bacon, *Novum Organum*

Time travel remains a science fiction staple, and through this medium just about everyone is familiar with the paradoxes involved. The intent here is not to prevent a workable blueprint for time travel but to create and test an analogy against the paradoxes inherent with the subject. By creating a plausible time travel scenario (that doesn't involve wormholes or flying on the edge of black holes) the tangle of logical inconsistencies involved with the subject can be unraveled and a logically sound definition of time created.

Let's begin by studying the paradoxes. The most recognizable of these problems is the "grandfather" paradox. The obvious issue here involves a traveler going "back" in time and killing his own grandfather, something that will prevent himself from ever having been born, which would, in turn, prevent the entire scenario from ever occurring.

A version of this paradox was featured in the 1980s classic film *Back to the Future*, starring a young Michael J. Fox as the protagonist Marty McFly. The plot, for anyone who hasn't seen the movie, involved the main character going back in time thirty years to the year 1954 where his own mother develops a crush on him. By inserting himself in time, the main character very nearly prevents his own parents from meeting, something that would have facilitated his own disappearance.

Of course, our hero does manage to get his parents back together, thus saving his own skin but altering history for the better in the process. (The final scene of *Back to the Future* features the main character engaging with a new and improved version of his family.) Of course, time travel is its own genre in sci-fi novels The plot usually involves the main characters being sent back in time for some spurious purpose, screwing something up, and then frantically trying to undo the effects of their actions. (That actually sounds a lot like what happened in *Back to the Future*.)

The best known of these novels is also the first. H. G. Wells's classic *The Time Machine* set a precedent for the genre that virtually every other time travel yarn follows. In these types of books, the time travelers always get into a machine of some kind that takes them back in time. This concept is all in good fun, but rife with logical inaccuracies. In his book *Timeline* Michael Crichton had one of his thinner characters, the sinister Doniger, explain away the concept of going back in time to change events by stating that it's highly unlikely that any one person could alter historical events. His point is that it's hard to alter history in your own time, so why should it be any easier to do so in the past? Even changing the outcome of a professional baseball game would prove to be nearly impossible.

Connie Willis is maybe the most widely read contemporary novelist to work in the time travel genre. Her books prove to be enormously entertaining, but all of them include the same error present in Crichton's novel and everywhere else. Time travel writers underestimate the ability of potential time travelers to affect the future with even a small interaction.

Consider this notion, oft repeated by Richard Dawkins: The male testes create billions of sperm cells every week. When a man impregnates a woman, only one of those sperm makes it to the egg. If a different sperm reached the egg, then a different combination of DNA would later be born. If a traveler did manage to go back in time and, say, accidentally bump into some man on the sidewalk, it's possible that this man would have his sperm jostled or a potential copulation with his wife would be delayed. Even if not now, then at a later date, that small interaction could change the time of his potential copulation.

This would almost certainly lead to a different sperm meeting the egg. So if Marty McFly so much as altered the day of his future father by even a few seconds, then the elder Mcfly and the soon-to-be Mrs. McFly may have done the under-the-covers-with-Barry-White-playing thing a little later or sooner than they originally intended, which would mean that a different Spermy McFly would reach the egg. This would then mean that Marty would never be born anyway, despite his best efforts.

In *Back to the Future*, Marty comes back to find a new and improved, but recognizable, family. He has the same brother and sister, but they are better dressed and more successful (and for some reason still in the family kitchen despite being older and having jobs).

Again, anything that Marty did in the past that would have altered the times of conception would have affected which sperm met the egg, and this would have led to different people being born. The same is true of any time travel tale. Any interaction with anyone would likely change the time of a potential copulation.

A time traveler might have an innocuous chat with someone at a fruit stand. That person might then be late coming home, and this alters a series of events so that this person later copulates with his wife seven seconds later than he would have in the original timeline. A different sperm hits the egg, a different child is born, and history is altered.

But this only involves macroevents. At another level, the time traveler would have to alter something. If the time traveler merely stood on the ground, then the molecules in the dirt or carpet would be slightly different than they would have been had there been no one standing there.

This is a difference that would show up in a future world, thus making the world visited by the time traveler slightly changed from the way it was originally. This may not alter macroevents but shows the impossibility of a time traveler engaging as a passive visitor to history.

Many sci-fi authors intuit this problem and work into the narrative some version of a time traveler who travels back in time, influences events, and then arrives back in the present to see that the present he had originally been in had already been influenced by his time-traveling self but in a way that was hidden from the time traveler until he came back to the present.

For example, one could go back in time, stop someone from shooting the president, then come back to the present and see that said president was never assassinated. Then the time traveler could sit with a wry smile every time this president was mentioned.

This answer to the time travel paradox still fails to satisfy even though it answers the basic contradiction of how a time traveler could go into the past without altering future events. The "present" that the time traveler originated in would have itself already been altered by his backward-in-time shenanigans.

The reason this is unsatisfying is because we have a problem of causality. We would have to posit an original timeline, somewhere and sometime, where events had not been altered in order to get to the original time traveler. That original time traveler would be trapped by all of the paradoxes already outlined in the above section. How, exactly, would the timeline have gotten started?

In order to explain, we'd have to theorize multiple universes, something that is not much different from merely saying, "God did it." Besides, the multiverse hypothesis still would not rescue us from the problem of finding an original universe where someone went back in time for the first time in order to alter the universe in such a way that he would later be able to live in that altered universe.

Time travel paradoxes can be maddening, and the intent here is not to try and poke holes in the fabricated space-time continuums issues of science fiction. Readers of sci-fi should not let a few scientific or philosophical problems get in the way of enjoying a story. (Statements like that redden the cheeks of a good many sci-fi fans, but oh well.)

This tangle of paradoxes and fallacies indicates that we are, profoundly and irrefutably, thinking about time in a fallacious way. The fourth-century theologian and wit St. Augustine famously stated that he knew what time was until asked to explain it. He finally concluded, "Time does not exist without some movement."

Think about this: a ruler measures distance, or space. A clock looks just like a ruler bent into a circle (and recall that both originated during medieval Europe's passion for quantification). Einstein recognized this with his conception of space-time. Here's a wingdinger of a question: What exactly is this circular ruler measuring? The best statement is probably to say that the clock measures movement, which is what we would call time.

This concept of time as a measurement of movement provides a way to understand physics. Movement, for example, did not exist prior to the Big Bang. All the particles in the universe existed in some microscopic little kernel (and man, was that sucker *dense*), but everything was so tightly packed that movement failed to exist. No movement = no time. The Big Bang cannot then be understood as an event on a timeline. There was no "before" the Big Bang, only an after.

Let's try this analogy, which might clear up any further paradoxes: Time is like particle and wave position. In this analogy, time is equated with whatever position the particles and waves in the universe, or any pocket of the universe, are in at any given moment sans movement. By reversing particle and wave positions to where they were at some point previously, one could, in actuality, go back in time.

This causes no paradoxes, because by reversing particle positions, one is causing forward movement to a backward position. So, picture a timeline where the particles and waves in the universe are all set in a certain way in 1950 (try to think in black and white), and then imagine those particles and waves shifting positions to, say, the year 2000.

In order to travel back to 1950 this would "merely" require moving the particles and waves back into the same position they were in during 1950. The particles and waves would be moved back to a previous state, but actions taken in this new state could lead to new outcomes that would in no way interfere with the past.

In fact, no one would even know that that 1950–2000 had already taken place before if everything, including neurons, was put back in the spot they were in 1950. But this 1950 state would be a set of particle and wave positions that would have taken place after the original. A new set of outcomes would be unproblematic for the logician since the changed particle and wave position itself would be an event on a line.

Now, if one wanted to time travel in this scenario, simply reverse the old H.G. Wells concept. In *The Time Machine*, and all the other sci-fi stories about time travel, a traveler gets into a machine and his molecules get tossed backward or forward on the timeline. Imagine a scenario where the time traveler gets in a machine where the purpose is not to send those molecules backward or forward but to preserve those molecules as they are while the rest of the universe reverts back to a previous state of particle and wave positions.

It is theoretically possible, not at this scale but at a smaller scale, to change particles and waves back to a previous state. Let's say we wanted to send ten molecules back to a previous state, say the way they were a decade ago. We'd have to calculate the odds as to where the particles would be based on their current state and then reverse their positions. Sure, there would be small mistakes made. Not all of the particles would get back to their exact positions, but in the aggregate the particles could arrive intact.

If this seems fishy, remember that we are always traveling in time, just forward at a certain rate. One cannot know, for example, that the rock in front of one's house will look tomorrow like it does today, and at a quantum level it likely won't. But in the aggregate, those particles will travel into the future in much the same way they always have.

We have to remember that as long as the changes in the universe would not affect the development of a human observer, from our perspective the changes are irrelevant. If we imagine a rock and trees as being placed back in their 1950 particle and wave position, it's likely that not every single quantum particle would be as it had been several decades prior. But what difference would that make to a human observer at the Newtonian level?

The analogy of time being like particle and wave position unravels the many paradoxes of time travel but also reveals the impossibility of creating any kind of "machine" that would act in accordance with this logic. Yet, the concept helps us to understand that time could exist prior to the Big Bang.

THE SINGULARITY MYTH

Medieval philosophers worked frequently between the concepts of universals and singulars, largely because this concept consumed Plato and because Aristotle disagreed with his master's position. (Remember that Augustine worked Plato into Christian dogma while Aquinas wove Aristotle into it.) We can now understand universals as phenomena for which philosophers have ready analogies. We can understand singularities as phenomena for which philosophers have no ready analogies.

The Big Bang falls into the singularity camp because there existed no movement and, therefore, no time prior to the Big Bang. In quantum calculations this is expressed as T = 0. Under the *new* new method, practitioners can begin with various propositions and then see whether those prepositions get strangled in their own contradictions. T = 0 is based on the concept of T = movement. Yet, as shown by the time travel paradoxes, a concept of time as a measurement of movement chokes itself.

Let's try T = PWP for time equals particle and wave position. This would conceive time as chopped up into small bits, let's say three-second "nows," which is about as long as it takes to process a conscious thought. (We are dealing with observers here after all.) This would put the present as a PWP with the past and future being PWP states assumed by probability.

Of course, a lot of movement can occur in three seconds, but we can consider that to be a part of a frozen particle and wave state. The position of a particle can be closely ascertained using probabilities into the near past or future with a high level of accuracy. This is about the best that can be done with an observer as part of the picture.

The advantage of such a system is that it makes the way we view the past as directly analogous, or consilient, with how we view the future. We see the future as a probable state of PWP, and those nearest to our current PWP are the easiest to predict. It's a safe bet that the current PWP of your living room will be similar to what it is in ten minutes but decisively less certain in ten million years. Likewise, we can predict the PWP of the living room for ten minutes ago much better than we can for ten million years.

This frees us from paradox but also lets the air out of some important analogies. Stephen Hawking (2001) has an analogy that trying to find a singular beginning to the universe is like trying to walk on the earth and find the end. The earth appears flat from your local position but gets round as you go.

The analogy never quite fits, as the question about the Big Bang singularity really involves a different type of question, a something-from-nothing paradox, and the question would more likely be, "Where did this round earth come from?" for which his analogy provides no answer.

This is not to disparage Hawking or anyone else. No theoretical graveyard overflows like the one outside the church of the something-from-nothing singularity. The T = PWP hypothesis stated here does not solve that particular problem but is intended to showcase the potential method of trying out different starting points until one can develop a coherent theory or analogy. This is the type of thing that Einstein did.

Now, according to the *new* new method, we must consider the possibility that our extrapolations are incorrect. The expanding universe concept and the Big Bang conclusion (or its cousins in the field of radiation theory and the like) have an ultimate conclusion based upon those extrapolations. But, this is based upon the conceit that the way light behaves now is directly analogous to how light behaved over thirteen billion years ago. That's a big assumption given that the conceit ultimately leads to its own contradiction.

The assumption that the universe's current activity is analogous to that of the past leads one to a singularity for which we have no analogy and, therefore, undercuts the very assumption that the entire edifice is built upon.

From the T = PWP position, a Big Bang singularity exists not as a fact but as one possibility. Obsessing over its specifics may be as fruitless as trying to work out equations to find a potential future for the universe, one that is rife with contradictions, based on the premise that the universe's laws 13.7 billion years from now will be directly analogous to those of today and then trying to work out the details of this single possibility. The Big Bang singularity may not be a problem but a statistical likelihood that, by virtue of its contradictions, may not be likely at all.

If this is the case, then our vibrating strings and elaborate equations start to look like the medieval attempts to prove God's existence with syllogistic logic. It can't be done, because God does not exist. The Big Bang probably did not either. We accept this logic elsewhere in physics. Quantum particles do not behave in a way that is analogous to how particles behave in the aggregate. Light sometimes behaves in a way that is analogous to a particle and sometimes to a wave. We cobble these analogies together into M-Theory and hope for the best.

So what did happen at the beginning? According to both MDR and T = PWP, it is not possible to know, only to assign odds to various theoretical propositions. The theory devoid of analogous contradictions is the most likely to be right. With T = PWP there would be potential periods of time where particles and waves existed and potential periods where they did not. We might assume one frame to be empty while the other contained something, such as energy, and then ask the question, "Where did that come from?"

This conception is consistent with Hawking's vision of the potential realities of the Big Bang, a position encapsulated by Paul Davies:

> As Hawking has emphasized, it is a mistake to think there is a single, well-defined cosmic history connecting the big bang to the present state of the universe. . . . Rather, there will be a multiplicity of possible histories, and which histories are included in the amalgam will depend on what we choose to measure today. "The histories of the universe depend on the precise question asked," Hawking said in a paper . . . with Thomas Hertog. . . . In other words, the existence of life and observers today has an effect on the past. "It leads to a profoundly different view of cosmology, and the relation between cause and effect," claims Hawking. (Davies, 2007, pp. 301–303)

This is not quite true; we observers do not affect the past any more than opening up the box containing Schrodinger's cat affects whether the cat lived or died in a past state. We are in a present box of consciousness; we assume that the boxes directly behind us and directly in front of us will act in a way that is analogous to our current physical reality. The odds are very good that these boxes have behaved, and will behave, in this way.

However, the further we get away from our current box, the less certain we can be of our analogies. And as we get back to a past that is so far distant that no observers existed, we have to extrapolate based on the evidence we compile and the analogies we create in our current boxes. If analogies are based on certain evidence, we create analogies that we extrapolate backward and assume that past or present boxes are either likely or unlikely.

Observers do not affect the stuff of the universe at all. The problem is that the universe doesn't look, taste, sound, or feel like anything. It is just stuff, and we can't imagine what it "really" looks, tastes, sounds, or feels like without an observer. For example, there must be a certain number of grains of sand on Earth. There would be a number if someone bothered to count them, but because we have not counted them, the number does not exist.

This is important to note because the idea that there is an ultimate consciousness with "right" answers and an ephemeral clock keeping "real" time remains oddly prevalent. The history of the universe had a single path, but, like the number of grains of sand, it remains inaccessible to us and, therefore, does not exist. Facts do not hang in the universe without observers because the facts are just things.

Take the death of Genghis Khan, for example. No one knows how he died. We know he must have died some way, but we have no evidence for how it specifically occurred and no evidence to help us prioritize the odds. We can assume he wasn't killed by aliens and that he didn't rise from the dead three days after being killed. Other than that, the particulars of his death exist only as probabilities, not as realities.

The reality of how he died is gone. The question brings us back to the "something from nothing" question. Under this system it is possible that we would go back to a first empty frame. The next frame would contain something, either a wave or particle. How did that get there? To begin, this exists only as one potentiality, and asking this question is not much different from asking how one's life would have been different if one had chosen to take up Ping-Pong as a profession at a young age.

There's no way to know because that decision would have spun off into a variety of random events that might have ended up with a car crash, a happy wedding, or any number of infinite possibilities, some more likely than others.

But in this single hypothetical past we could take our pick from string theory or other potential "something from nothing" theories to explain. There's no shortage of these theories, and they are nicely encapsulated in a book edited by Robert Lawrence Kuhn and John Leslie (2013), titled *The Mystery of Existence: Why Is There Anything at All?*

Ultimately, the question posed by the title is probably moot. In a universe with something, the most likely answer is that each PWP frame in the past contained an uncreated something. In a universe with nothing, the most likely answer would be that each PWP frame in the past contained nothing. But in that universe, no one would be around to figure out the probabilities.

The analogies we create from our current PWP would be the most likely to fit frames that are closest to our current perspective and less likely the further we get. Imagine an infinite stream of letters: only where the letters actually spell a coherent paragraph would the conditions be right to create a sentient observer.

That observer might discover that she stands on a series of letters that operate under grammatical laws and make sense. If she tried to extrapolate her understanding onto the infinite chain of nonsense she would find herself frustrated once she got away from the cozy region of her coherent paragraph.

Physicists who fret over the fact that the universe would dissolve into chaos were it not "fine-tuned" to so many decimal points put too much faith in their numbers. Their navel-gazing is no more significant than an English professor marveling that if the laws of English grammar were any different, then Shakespeare's play might not look exactly as they do. The odds against anything are astronomically high if one adds enough factors.

For example, the odds that you will have breakfast are good. The odds that you will have eggs are pretty good. The odds that you will have eggs and bacon are not bad. But the odds that you will have precisely 0.356789 pounds of eggs, 0.23454 pounds of bacon, and 1.23432 liters of orange juice are not so good. Yet numbers like this occur everywhere and at all times, without anyone remarking on their near impossibility.

One could speculate that a god must have steered all these events. One could likewise speculate that all the people who don't exist as a result of chance meetings and sperm-egg connections must exist in a polycosm or that all probable weights of your breakfast exist in alternative realms, but that gets goofy in a hurry, doesn't it?

We have no hard evidence for either conjecture. The problem with the fine-tuning conjecture is that it is analogous to every single event if we add enough factors, and for those ordinary events we do not posit extraordinary causes. By analogy, why should we post extraordinary causes for the existence of the universe?

T = PWP clarifies the more confusing aspects of relativity theory. Everyone is familiar with the mind-warping concept of time dilation, in which clocks slow down the faster the vessel carrying them moves. According to relativity theory, someone traveling at close to light speed in a ship would experience one day while people on Earth experienced two years. (This concept varies widely depending on how close one gets to light speed.)

How can this be possible? Someone on the ship watches the hands on a clock circle twenty-four times while a full two years passes for someone on the ground? The confusion comes only when time is assumed to be the measurement of movement instead of the frozen and framed positions of particles and waves. Movement seems like it should not occur at different rates because our very concept of movement developed in minds that all travel at the same speed.

Instead, picture a line with the observer at rest represented at the bottom. His frames of PWP are divided into relatively small sections. For purpose of simplicity, let's imagine a four-year timeline divided into 730 boxes. Now, imagine two larger boxes on top of the timeline that take up the entirety of the timeline. This represents the observer moving at nearly light speed. The two men would move through their boxes and then, if the observer on the ship slowed down, meet at the end of the timeline. The faster one moves, the bigger one's particle-and-wave-experience box becomes at a mathematically discernible rate.

The observer on the ship moved only through two boxes of particle position, or two days, while the observer on the ground moved through 730 boxes of particle position. The observer on the ship has particles and waves in his body that are only two days older, while the observer on the ground has particles and waves which are 730 days older. If this still sounds mind-warping, then remember that we are talking about space-time. Speed alters the way we experience distance, and it also alters how we experience time.

Imagine a bizarre universe where everyone walks at the same rate. Someone would assume in this universe that it might take fifteen minutes for a friend to walk to her house. If that friend showed up in seven minutes, by discovering running, then these people might be shocked to discover that distance could be experienced differently at different speeds.

Thus we see that the concept of T = PWP does everything that T = movement accomplishes, but also clears up some logical grey areas and brings the theory more closely in alignment with the probabilistic properties of quantum mechanics. The beginning of the universe is inaccessible; what we have are boxes to be judged on probability based on the analogies we create from whatever evidence we choose to study. And the Big Bang singularity cannot be right since its conclusion undercuts the premise that our analogies hold through the entire connecting chain, up until the last link, when they do not.

CONSCIOUSNESS

Having established the preceding principles, the analogies can be superimposed on the problem of consciousness, sometimes known as the mind-body problem. The problem involves connecting the stuff of the brain with the abstract notions of thought that it presents. This problem is not as complex as the Big Bang singularity question since it does not have to account for outside energy into the system, and an analogy from MDR can aid understanding.

We know that thought is based on the stuff of the brain because of the impact of injury on the brain itself. If a scalpel removes a section of the brain, or if blunt-force trauma damages a particular section of the brain, then this manifests itself in the behavior of an individual. This happens without exception.

Someone who has her frontal lobes removed will act differently than she did before the operation, and people who die no longer exhibit personalities since the brain itself has ceased to operate. No ghosts, spirits, or souls exist independent of the machinery.

The human brain develops in accordance with environmental stimulus. The medieval Arabic Aristotelian Avicenna (980–1037) developed the famous "flying man" thought experiment. He wondered what would happen to a fully developed man suddenly suspended in the air, blindfolded, and separated from sensation. He concluded he would still have an intellect, thoughts, and, therefore, a soul separate from the body.

The obvious fallacy with his thinking is that he imagined a flying *man* whose consciousness and thoughts had already been shaped by sensory experience. He should have imagined a flying baby. What kind of consciousness would a baby in a sensory deprivation chamber develop?

One hates to subject even a hypothetical baby to this experience since we can guess the result. His neurons would not develop, his tongue would never shape words, and his mind would be starved of the fuel it would need to create coherent thoughts.

We can imagine the shaky steps of a toddler as being similar to the shaky fingers of an adult working for the first time on a piano keyboard.

The body needs to take in outside sensory data in order to shape the brain and its thoughts. The total of this is called consciousness. But where do thoughts or the will to generate the body's movements come from? We might view this as a feedback loop between the environment and the brain. Certain situations create certain types of feedback in the brain so that most actions take place unconsciously.

The brain connects patterns of experience so that when one encounters a situation that is analogous to a situation encountered before, a pattern driven by outside sensory data lights up. If the previous experience had been positive, then the same action might prove possible in this analogy. If not, then a new action might be required. This cognition is called "critical thought and analysis."

Instinctual animal behavior might be described as rudimentary analogy-making encoded genetically. Animals act in certain ways because their genes have encountered analogous situations in the past. This is why altering an animal's habitat can lead rapidly to extinction, since the genes have analogies encoded that only work in environments that are analogous to previous genes.

For humans, analogy-making becomes more nuanced through experience and education. The environment lights up certain pathways created through experience, and these analogies direct action. Analogical thought predated language, and so the evolution of the voice box, credited by Jared Diamond and others for pushing *Homo sapiens* into a cultural "Great Leap Forward," likely gave voice to analogy-making. This allowed analogies to be shared and shaped in deeper ways, a process that continues to be facilitated by the development of technology.

Cognition is not the top-down controller of analysis but the bottom-up pinnacle of analogous thinking. We need not worry about where a thought generates because the energy comes from outside the system and is directly analogous to things that we understand. The richer our experiences, the more nuanced our analogies and the more coherent our thoughts. Thinking of a core "I" inside of us is no more useful than William Paley's conjuring of a "watchmaker" who created the universe. We are the sum of bottom-up analogical structures, a sentient sequence of analogies reacting to situations using our experience.

KEY POINTS

Approaching age-old problems in physics and philosophy as crises of analogy provides a new way forward for thinkers. These problems are not like mathematical problems in that they cannot be solved with single answers.

Removing movement from the theoretical picture brings us closer to understanding the true place of statistical analysis in the big picture of theoretical physics.

The notion of a Big Bang singularity is presupposed on the idea that the universe's "laws" from almost fourteen billion years ago are analogous to the laws we have today. We must presuppose a very long chain of analogy into the past until we get to the point when we encounter something for which we have no analogy at all—a singularity. This last box draws into question the entire enterprise.

Trying to predict the past is directly analogous to trying to predict the future.

We can no more predict the distant past than we can the distant future. Our assumptions and understandings are all based upon analogies we've formed in our brief "box" of consciousness.

Everything is simplified by redefining time as a statement of particle and wave position, as opposed to a measurement of movement.

Consciousness, too, can be seen as the process of analogy-building. Instincts are simplistic analogies encoded into the DNA. Experience and education build on analogies, but our consciousness must receive constant energy from the outside world. Without this outside sensory evidence, we have no consciousness.

Conclusion: The *New* New Atlantis

> There will be found no doubt, when my history and tables of discovery are read, some things in the experiments themselves that are not quite certain, or perhaps that are quite false; which may make a man think that the foundations and principles upon which my discoveries rest are false and doubtful. But this is of no consequence; for such things must needs happen at first. It is only like the occurrence in a written or printed page of a letter or two mistaken or misplaced; which does not much hinder the reader, because such errors are easily corrected by the sense.
>
> —Francis Bacon, *Novum Organum*

> Every research mathematician is aware that, at any moment, suddenly and unpredictably, the problem they are working on may turn out to require ideas from some apparently unrelated area. Indeed, new research often combines areas.
>
> —Ian Stewart, *Visions of Infinity: The Great Mathematical Problems*

Take Einstein—he had to be able to think analogically, see new truths, and then be able to translate those truths into mathematics. Moreover, he had to live in a part of the world that possessed mathematical journals. Other examples of geniuses who have developed analogies from one field and then applied them to another abound in the history of science.

Thomas Kuhn (1996) even observed, "Almost always the men who achieve these fundamental inventions of a new paradigm have been either very young or very new to the field whose paradigm they change" (p. 90). Scientists or thinkers of a certain generation tend to have their thinking on certain questions calcified; only when someone comes in with a fresh perspective or novel background to look at a problem are breakthroughs made.

Isaac Newton made his leaps by ignoring the stagnant Cambridge curriculum; instead, he began with big questions about the universe and started with first principles to answer them. Faraday was an interested amateur with a gift for devising experiments. Einstein thought of novel questions and then approached the process of answering them from first principles, almost totally ignoring the scientific and mathematical establishment in his quest. Jared Diamond approached history from the perspective of a biologist and, beginning with principles of evolution, set out to answer some of history's biggest questions.

August Kekulé, a pioneer in organic chemistry, was originally trained in architecture before entering chemistry, and Peter Watson notes that Kekulé "later . . . argued that his architectural training (such as it was) had helped him to think in pictures—and this played a vital role when he came to identify the structure of carbon compounds" (2010, p. 278). Likewise, Masha Gessen writes that the modern genius Grigory Perelman proved the Poincaré conjecture by "applying Alexandrov spaces to Geometrization" (2009, p. 122) and uniting various concepts from different mathematical backgrounds.

If it is the case that great breakthroughs are made by people approaching fields from another perspective (and when Thomas Kuhn, E. O. Wilson, and Douglas Hofstadter all make a similar point, we should pay extra attention), then why leave this process up to the fates? As it stands now, great breakthroughs happened by chance, only when someone well versed (even obsessive) in one field of study develops a passionate obsession for topics in another field.

We have to wait for the rare person who can engage in this process and then we all too often *also* require that this individual be able to translate the analogical findings into mathematical terms.

These are too many necessary conditions for fire. With this many conditions, we should not marvel that we only occasionally get a Newton or Einstein. We can lessen these conditions. First of all, it is possible to train people to think via analogy. The notion here would be to involve individuals in metacognition by waking people up to the force of analogical thinking in their lives.

Secondly, higher-level thought needs to be unchained from the anchor of mathematics.

As mentioned earlier, all too often in history breakthroughs have occurred when someone studies deeply in one field, finds patterns there, and then superimposes those patterns on a new field of study.

This type of thinker sees novel analogies in one field of study that can then be applied to problems in another field. The process of achieving these breakthroughs is rare, largely because it is randomized. Only rarely will such obsessive polymaths occur out in nature.

But genius can and should be domesticated. Instead of waiting for the rare polymath to emerge, we should redefine what "new knowledge" means.

What if, instead of training people in specific fields of study with the goal of creating both experts and knowledge producers in that field, we created a new field for "analogy and consilience"? Practitioners would be professional analogy hunters who looked for hypotheses that are consistent across fields. One set of facts might produce a pattern that also holds true for analogous situations.

The analogy would be the new information. Again, lattice theory, for example, might contain analogies that could bring useful insights into the way in which the brain forms neuronal connections. We cannot expect that someone will get a PhD in both deep mathematics and neuroscience. What we can do is turn the hunt for analogy itself into a field and define these analogies as "new" knowledge. This is the *new* new method.

One point too large to miss about Francis Bacon's *Novum Organum* is that the full impact of Bacon's ideas was not felt until they were inculcated in the educational system, first in universities in Germany and then in medical schools in France. One might describe this process as the institutionalization of scientific genius, in the sense that Bacon saw that experimentation represented a forward shift in the way that thinkers addressed knowledge.

The fact that it took about 150 years for his ideas to spread and change the gestalt to the point where schools picked up on his concepts is sad. I am a lowly schoolteacher, and, even worse, I have a doctorate in education (not even a doctorate in philosophy!). Therefore, on the intellectual food chain, I have the equivalent rank of a small, blind, badly injured rodent.

Yet my background in education allows me to see the absolute importance of building the *new* new method into the educational system, not as replacement to Bacon's core philosophy (remember, this book is intended as a sequel) but alongside the already existing research institutions.

Bacon's novella, *New Atlantis*, gave fictional life to his ideas by positing a utopic vision of what a world, or an island in this case, where philosophers embraced his ideas would look like. I propose here a New Atlantis for education and will, in as few words as possible, restate the basic philosophy presented in these earlier pages so as to establish the logical links that lead to the conclusion.

The *new* new method views scientific and mathematical theories as arbitrary explanations of phenomena. Yes, even relativity theory must be now understood as a series of arbitrarily made insights and analogies that match with actual activities in the universe all or most of the time. (I reiterate and stand by my point that light probably acts in a messier way than relativity predicts.) Neal Stephenson (2003) explained that mathematicians are Platonists, believing that they are uncovering truths rather than creating languages.

The *new* new method views all thinking as analogical at its core. Mathematical and syllogistic forms of logic are secondary, and the current place of mathematics in the educational system may be viewed as analogous to the place that theology and Latin had in the medieval educational system. Latin and theology were considered to be of great importance when the emphasis was on rote memorization of classical texts and church dogmas.

When the emphasis switched to Aristotelian reasoning, those subjects became less important. Now that the centrality of analogical thinking is established, the importance of mathematics should also shift.

Analogical thinking can be judged by medieval forms of reasoning. Analogies should be created and judged by their level of consilience. Does an analogy fit a variety of different facts and situations? If it does, it can be said to be consilient and, therefore, strong. Again, Darwinian analogies fit every single thing one encounters, showing what a powerful theory Darwinism is.

Historically, breakthroughs in the sciences occur whenever someone with a passion for content in two fields manages to bring a new set of analogies (derived from a sometimes far-distant field) to problems in another field. This random process can be domesticated, but the current university system, derived from Bacon's original principles, does not facilitate this historical process.

Too many so-called educational reformers actually call only for the intensification of the existing system. The hunt for analogies itself must be turned into an educational system.

Teachers, at the secondary level, engage in a field of teaching that is separate from education and from content areas, but the field of teaching comprises elements of both. The "new knowledge" produced by teachers is a synthesis of educational methods and content-based research.

Teachers who think in this way will develop analogies for the education of students, and the purpose of content will be to create more nuanced analogies (thus avoiding the naïve analogies that plague thinking).

We must confront the possibility that our very language is unsuitable for the tasks at hand. Philosophers, currently, are trying to adapt human languages that were originally intended for intratribal communication to questions that might be beyond the capacity of language to explain. We should expect that different languages will have different levels of descriptive power.

Will Mandarin, for example, prove to be the language by which the best neuroscience is conducted? After all, the brain may store information in a way that is more analogous to the pictographic languages of China than to the phonetic languages of the West.

Or, is it possible that all human languages will prove to be ultimately inadequate to the task of describing analogies?

What if we developed, in time, forms of artificial intelligence (AI) that could rapidly create languages specifically suited to provide analogies for the singularities? Would the AIs develop languages that work in ways we could not fully comprehend? Would this mean that the AIs would develop theories that are inaccessible to the human mind yet still internally structured and strong?

One might notice that very little has been said here about the impact of modern technology. That's because the irony of the age may end up being that the very objects that make information readily available to so many people are at the same time destroying the attention span that makes it possible to understand and synthesize that information.

It is possible that technology provided a small pocket in history where information became readily available at the same time that people had the attention span to think about it deeply. Technology may giveth access to information but, in due time, taketh away the attention span.

Perhaps computer programmers will create AIs that operate by creating and judging analogies. Would this haphazard process lead to a series of organic and novel connections? Would an analogy machine eventually "wake up" and become sentient? What would it say?

Maybe *Novum Organum III* will be written by an AI.

For practitioners of the *new* new method, an environment including a formalized system of study may not be the best fit. We might consider this Richard Rhodes quote about the European system of higher education, prevalent in pre–World War II Europe, which led to so much original physics:

> Physics students at that time wandered Europe in search of exceptional masters much as their forebears in scholarship and craft had done since medieval days. Universities in Germany were institutions of the state; a professor was a salaried civil servant who also collected fees directly from his students for the course he chose to give. . . . If someone whose specialty you wished to learn taught at Munich, you went to Munich; if at Gottingen, you went to Gottingen. Science grew out of the craft tradition in any case; in the first third of the twentieth century it retained—and to some extent still retains—an informal system of mastery and apprenticeship over which was laid the more recent system of the European graduate school. (2012, p. 16)

The hunt for analogies as knowledge might require a looser organization so as to not stifle creativity. Bacon's followers in the Royal Society fancied themselves to be the bees he called for. They gathered information from the world and put it to good use, just like bees did when building hives. The Royal Society created the Latin motto of *Nullius in Verba,* which encapsulated their ideas. It meant "on the words of no one" and reminded the new natural philosophers that they should not take any preposition or theory, as the hated scholastics had done, merely out of undue respect for authority.

Practitioners of the *new* new method, however, must be polymaths who search in the dark corners of esoteric subjects looking for knowledge that can be molded into novel analogies.

Perhaps, then, I might suggest a new motto: *Quisque in Verba.*

On the words of everyone.

References

Al-Khalili, Jim. (2011). *The House of Wisdom: How Arabic Science Saved Ancient Knowledge and Gave Us the Renaissance*. New York: Penguin.
Aquinas, Thomas. (1973). The Summa Logica. In Arthur Hyman and James Jerome Walsh (Eds.), *Philosophy in the Middle Ages: The Christian, Islamic, and Jewish Traditions* (2nd ed.) (pp. 503–581). Indianapolis: Hacket.
Bacon, Francis. (1994). The Great Instauration and Novum Organum. In Edwin A. Burtt (Ed.), *The English Philosophers from Bacon to Mill*. New York: The Modern Library.
Bacon, Francis. (2008). *The Major Works: Including New Atlantis and the Essays*. Oxford: Oxford University Press.
Bacon, Roger. (1973). The Opus Majus. In Arthur Hyman and James Jerome Walsh (Eds.), *Philosophy in the Middle Ages: The Christian, Islamic, and Jewish Traditions* (2nd ed.) (pp. 474–480). Indianapolis: Hacket.
Baker, Nicholson. (2013, September) Wrong Answer: The Case against Algebra II. *Harper's, 327*(1960), 31–39.
Ball, Phillip. (2010). Making Stuff: From Bacon to Bakelite. In Bill Bryson (Ed.), *Seeing Further: The Story of Science, Discovery and the Genius of the Royal Society*. New York: William Morrow.
Barbour, Julian. (2001). *The End of Time: The Next Revolution in Physics*. Oxford: Oxford University Press.
Bauer, Susan Wise. (2013). *The History of the Renaissance World: From the Rediscovery of Aristotle to the Conquest of Constantinople*. New York: W.W. Norton..
Bourke, Vernon J. (Ed.) (1974). *The Essential Augustine*. Indianapolis: Hacket.
Bowen, Catherine Drinker. (1963). *The Temper of a Man*. Boston: Little, Brown.
Buckingham, Will, Burnham, Douglas, Hill, Clive, King, Peter J., Marenbon, John, and Weeks, Marcus. *The Philosophy Book: Big Ideas Simply Explained*. London: DK.
Buridan, John. (1973). Questions on Aristotle's Metaphyscis. In Arthur Hyman and James Jerome Walsh (Eds.), *Philosophy in the Middle Ages: The Christian, Islamic, and Jewish Traditions* (2nd ed.). Indianapolis: Hacket.
Burke, James. (1985). *The Day the Universe Changed*. Boston: Little, Brown.
Canton, Neil, & Bob Gale (Producers), & Zemeckis, Robert (Director). (1985). *Back to the Future* [Motion picture]. United States: Amblin Entertainment/Universal Pictures.
Crichton, Michael. (1999). *Timeline*. New York: Ballantine.
Cohen, Marc S., Curd, Patricia, and Reeve, C. D. C. (2000). *Readings in Ancient Greek Philosophy: From Thales to Aristotle* (2nd Ed.). Indianapolis: Hacket.
Crosby, Alfred W. (1997). *The Measure of Reality: Quantification and Western Society 1250–1600*. Cambridge: Cambridge University Press.
Crosby, Alfred W. (2002). *Throwing Fire: Projective Technology through History*. Cambridge: Cambridge University Press.
Davies, Paul. (2007, June 30). The Flexi-Laws of Physics. *New Scientist*, pp. 301–303.
Dawkins, Richard. (2009). *The Greatest Show on Earth: Evidence for Evolution*. New York: Free Press.
Diamond, Jared. (1992). *The Third Chimpanzee: The Evolution and Future of the Human Animal*. New York: Harper Perennial.
Diamond, Jared. (1999). *Guns, Germs, and Steel: The Fates of Human Societies*. New York: W.W. Norton.

Edwards, Chris. (2011). Stephen Hawking's Other Controversial Theory. *Skeptic, 16*(3), 38–40.

Edwards, Chris. (2012). *Teaching Genius: Redefining Education with Lessons from Science and Philosophy*. Lanham, MD: Rowman & Littlefield Education.

Edwards, Chris. (2013). What's It Like? The Science of Scientific Analogies. *eSkeptic*. Retreived from http://www.skeptic.com/eskeptic/13–10–02/.

Fadiman, Anne. (1997). *The Spirit Catches You and You Fall Down: A Hmong Child, Her American Doctors, and the Collision of Two Cultures*. New York: Farrar, Straus, and Giroux.

Fenby, Jonathan. (2007). *China's Imperial Dynasties: 1600 BC–AD 1912*. New York: Metro.

Galileo. (1998). The Assayer. In Peter Machamer (Ed.), *The Cambridge Companion to Galileo* (pp. 64–65). Cambridge: Cambridge University Press.

Gessen, Masha. (2009). *Perfect Rigor: A Genius + the Mathematical Breakthrough of the Century*. Boston: Houghton Mifflin Harcourt.

Gibbon, Edward. (2003). *The Decline and Fall of the Roman Empire*. New York: The Modern Library. (Original work published 1776)

Gleick, James. (1992). *Genius: The Life and Science of Richard Feynman*. New York: Vintage Books.

Gleiser, Marcelo. (2010). *A Tear at the Edge of Creation: A Radical New Vision for Life in an Imperfect Universe*. New York: Free Press.

Goldstein, Rebecca Newberger. (2010). What's in a Name? Rivalries and the Birth of Modern Science. In Bill Bryson (Ed.), *Seeing Further: The Story of Science, Discovery and the Genius of the Royal Society* (pp. 107–129). New York: William Morrow.

Grayling, A. C. (2005). *Descartes: The Life and Times of a Genius*. New York: Walker.

Greenblatt, Stephen. (2011). *The Swerve: How the World Became Modern*. New York: W.W. Norton.

Gribbin, John. (1996). *Schrodinger's Kittens and the Search for Reality: Solving the Quantum Mysteries*. New York: Back Bay Books.

Gribbin, John. (2002). *The Scientists: A History of Science Told Through the Lives of Its Greatest Inventors*. New York: Random House.

Grosseteste, Robert. (1973). On Light. In *Philosophy in the Middle Ages: The Christian, Islamic, and Jewish Traditions* (2nd ed.) (pp. 474–480). Indianapolis: Hacket.

Hawking, Stephen. (1998). *A Brief History of Time: The Updated and Expanded Tenth Anniversary Edition*. New York: Bantam Books.

Hawking, Stephen. (2001). *The Universe in a Nutshell*. New York: Bantam.

Hawking, Stephen, and Mlodinow, Leonard. (2010). *The Grand Design*. New York: Bantam Books.

Hecht, Jennifer Michael. (2003). *Doubt: A History*. San Francisco: HarperCollins.

Hofstadter, Douglas. (2007). *I Am a Strange Loop*. New York: Perseus.

Hofstadter, Douglas, and Sander, Emmanuel. (2013). *Surfaces and Essences: Analogy as the Fuel and Fire of Thinking*. New York: Basic Books.

Inwood, Stephen. (2004). *The Forgotten Genius: The Biography of Robert Hooke, 1635–1703*. New York: Macadam Cage.

Kaye, Sharon M. (2008). *Medieval Philosophy*. Oxford: OneWorld.

Keay, John. (2009). *China: A History*. New York: Basic Books.

Kirsch, Joanathan. (2004). *God against the Gods: The History of the War between Monotheism and Polytheism*. New York: Viking.

Kuhn, Robert Lawrence, and Leslie, John. (2013). *The Mystery of Existence: Why Is There Anything at All?* Malden, MA: Wiley-Blackwell.

Kuhn, Thomas. (1996). *The Structure of Scientific Revolutions* (3rd ed.) Chicago: University of Chicago Press.

Lakoff, George, and Johnson, Mark. (1980). *Metaphors We Live By*. Chicago: University of Chicago Press.

Lapidus, Ira M. (2002). *A History of Islamic Societies* (2nd ed.). Cambridge: Cambridge University Press.

Lindsey, Jack. (2009). *Blast-Power and Ballistic: Concepts of Force and Energy in the Ancient World*. New York: Barnes & Noble. (Originally published 1974)

Loewenstein, Werner R. (2013). *Physics in Mind: A Quantum View of the Brain*. New York: Basic Books.

MacCulloch, Diarmaid. (2009). *Christianity: The First Three Thousand Years*. New York: Viking.

Machamer, Peter (Ed.). (1998). *The Cambridge Companion to Galileo*. Cambridge: Cambridge University Press.

Pagden, Anthony. (2008). *Worlds at War: The 2,500 Year Struggle between East and West*. New York: Random House.

Pendegrast, Mark. (1999). *Uncommon Grounds: The History of Coffee and How It Transformed Our World*. New York: Basic Books.

Plato. (2001). *The Apology*. (Benjamin Jowett, Ed.) The Harvard Classics (Millennium Edition). Norwalk, CT: Easton Press.

Reston, James. (2005). *Dogs of God: Columbus, the Inquisition, and the Defeat of the Moors*. New York: Random House.

Reynolds, Joshua. (1995). Discourse on art (Originally published 1776). In Issac Kramnick (Ed.), *The Portable Enlightenment Reader*. Viking Portable Library. New York: Penguin.

Rhodes, Richard. (2012). *The Making of the Atomic Bomb*. New York: Simon & Schuster.

Robinson, Daniel N. (2004). *The Great Ideas of Philosophy Course Guidebook* (2nd ed.). Chantilly, VA: The Teaching Company.

Roberts, J. M. (2013). *The History of the World* (6th ed.) Oxford: Oxford University Press.

Rubenstein, Richard E. (2003). *Aristotle's Children: How Christians, Muslims, and Jews Rediscovered Ancient Wisdom and Illuminated the Middle Ages*. New York: Harvest Books.

Runciman, Steven. (1990). *The Fall of Constantinople*. Cambridge: Canto.

Russell, Bertrand. (1945). *A History of Western Philosophy*. New York: Simon & Schuster.

Seife, Charles. (2000). *Zero: The Biography of a Dangerous Idea*. New York: Penguin.

Snyder, Laura J. (2011). *The Philosophical Breakfast Club: Four Remarkable Friends Who Transformed Science and Changed the World*. New York: Broadway Books.

Stephenson, Neal. (2003). *Quicksilver*. The Baroque Cycle No. 1. New York: William Morrow.

Stewart, Ian. (2013). *Visions of Infinity: The Great Mathematical Problems*. New York: Basic Books.

Stone, I. F. (1989). *The Trial of Socrates*. New York: Anchor Books.

Stott, Rebecca. (2012). *Darwin's Ghosts: The Secret History of Evolution*. New York: Spiegel & Grau.

Susskind, Leonard. (2008). *The Black Hole War: My Battle with Stephen Hawking to Make the World Safer for Quantum Mechanics*. New York: Little, Brown & Company.

Temple, Robert. (2007). *The Genius of China: 3,000 Years of Science, Discovery, and Invention*. London: Andre Deutsch.

Watson, Peter. (2005). *Ideas: A History of Thought and Invention, from Fire to Freud*. New York: HarperCollins.

Watson, Peter. (2010). *The German Genius*. New York: HarperCollins.

Whyte, Jamie. (2004). *Crimes against Logic: Exposing the Bogus Arguments of Politicians, Priests, Journalists, and Other Serial Offenders*. New York: McGraw-Hill.

William of Ockham. (1973a). Commentary on the Sentences. In Arthur Hyman and James Jerome Walsh (Eds.), *Philosophy in the Middle Ages: The Christian, Islamic, and Jewish Traditions* (2nd ed.) (pp. 662–686). Indianapolis: Hacket.

William of Ockham. (1973b). De Successivis. In Arthur Hyman and James Jerome Walsh (Eds.), *Philosophy in the Middle Ages: The Christian, Islamic, and Jewish Traditions* (2nd ed.) (pp. 686–689). Indianapolis: Hacket.

William of Ockham. (1973c). Seven Quodlibets. In Arthur Hyman and James Jerome Walsh (Eds.), *Philosophy in the Middle Ages: The Christian, Islamic, and Jewish Traditions* (2nd ed.) (pp. 690–693). Indianapolis: Hacket.

William of Ockham. (1973d). Summa totius logicae. In Arthur Hyman and James Jerome Walsh (Eds.), *Philosophy in the Middle Ages: The Christian, Islamic, and Jewish Traditions* (2nd ed.) (pp. 653–662). Indianapolis: Hacket.

Wilson, Edward O. (1998). *Consilience: The Unity of Knowledge*. New York: Vintage Books.

Wilson, Edward O. (2013, April 5). Great scientist does not equal good at math: E. O. Wilson shares a secret: Discoveries emerge from ideas, not number crunching. *Wall Street Journal*.

Wittgenstein, Ludwig. (2009). *Major Works: Selected Philosophical Writings*. New York: Harper Perennial.

Woodfin, Rupert, and Groves, Judy. (2002). *Introducing Aristotle*. Cambridge: Icon Books.

About the Author

Chris Edwards, EdD, is the author of three books of philosophy, including *Teaching Genius: Redefining Education with Lessons from Science and Philosophy*, and is a frequent contributor on the topics of philosophy, law, logic, theoretical physics, and education to the science and philosophy journals *Skeptic* and *Free Inquiry*. His original connect-the-dots teaching methodology has been published by the National Council for Social Studies. A twelve-year veteran of the classroom, he teaches world history at a public high school in the Midwest.

CPSIA information can be obtained at www.ICGtesting.com
Printed in the USA
BVOW01s1056030414

349580BV00003B/4/P